A NATURALIST'S GUIDE TO THE

SNAKES
OF
SOUTHEAST ASIA

including Malaysia, Singapore, Thailand, Myanmar, Borneo, Sumatra, Java and Bali

Indraneil Das

T0163120

JOHN BEAUFOY PUBLISHING

This edition published in the United Kingdom in 2021 by John Beaufoy Publishing
11 Blenheim Court, 316 Woodstock Road, Oxford OX2 7NS, England
www.johnbeaufoy.com

10 9 8 7 6 5 4 3 2 1

ISBN 978-1-913679-09-5

Photo captions
Front cover: *main image* Blue Coral Snake; *bottom left* Gumprecht's Pit Viper; *bottom centre* White-bellied Rat Snake; *bottom right* Sunbeam Snake, all © Indraneil Das.
Back cover: Large-eyed Pit Viper © Gernot Vogel.
Title page: Golden-bellied Reed Snake © Evan Quah. **Contents page:** Cambodian Puff-faced Water Snake © Nikolay Poyarkov.

Dedication
Nothing would have happened without the support of the folks at home: my wife, Genevieve V.A. Gee, and son, Rahul Das. To them, I dedicate this book.

Edited and designed by D & N Publishing, Baydon, Wiltshire, UK

Printed and bound in Malaysia by Times Offset (M) Sdn. Bhd.

·CONTENTS·

Introduction

Snakes form one of the major components of vertebrate fauna of Southeast Asia. They feature prominently in folklore, mythology and other belief systems of the indigenous people of the region, and are of ecological and conservation value, some species supporting significant (albeit often illegal) economic activities (primarily, the snake-skin trade, but also sale of meat and other body parts that purportedly have medicinal properties). They fascinate city-dwellers as much as they engage rural folks, but often suffer prejudices, perhaps borne out of fear, ignorance, and religious and other early teachings.

As of 2020, a total of 444 species of snake have been recorded from the region. This guide, now in its third edition, describes and illustrates 265 species (or nearly 60 per cent) of the snake fauna of Southeast Asia (here including Myanmar, Thailand, Peninsular Malaysia and Singapore, and within the insular region, Sumatra, Borneo, Java and Bali, as well as the smaller islands and archipelago systems of the Greater Sundas). All other regions are termed extralimital in this work (although the natural distribution of species that spill over political boundaries of these countries may not necessarily be biologically extralimital).

The aim of this work is to permit rapid field identification of each species covered via descriptive text and one or more live photographs. In particular, details of colour and form are noted in the descriptions. A few species in this work have never been illustrated in a publication before, and for all I have made an effort to use previously unpublished images. In several species, accurate identification in the field is not possible without detailed scale counts or other morphological examination, for which users will have to refer to more technical works (see Further Reading, p. 170).

Habitats

Southeast Asia is home to a remarkable diversity of snake life, due in part to the habitat conditions. Stretching from Myanmar to the islands of the Sundas, the habitat range encompasses the dry deciduous and subtropical forests of Myanmar that lie to the south of the western outliers of the Himalayas, the vast relatively low-lying areas of southern Thailand and the Malay Peninsula, and the often isolated mountain massifs of northern Myanmar, eastern Thailand and Sundaland; the southern regions relatively aseasonal, albeit with more wet periods, especially during the passage of the winter monsoons.

Natural habitats include lowland and hill dipterocarp forests that may reach subalpine limits, as in Gunung Kinabalu (4,095m asl) in northern Borneo; Hkakabo Razi (5,881m asl) in northern Myanmar and Doi Inthanon (2,576m asl) in northern Thailand. Also remarkable is the presence of more specialized habitats, including *kerangas* (Bornean heath) and vast tracks of forests associated with blackwater habitats that, because they are highly acidic, present important challenges for their biodiversity. Coastal habitats round up snake habitats in our region, and include mangrove swamps, beach forest habitats, shallow coastal seas and coral reefs.

Mangrove forests, Pulau Selurong, Brunei Darussalam

SNAKE IDENTIFICATION

Many species of snake can be told apart using details of their coloration, including both colour and pattern observed on their scales and, sometimes, underlying skin. Other features that are important for snake identification include the shape of their heads, robustness of their bodies, and relative length of their tails. The nature of scales present on different parts of the body is also instructive; they may be smooth, weakly keeled or bear multiple, sharp keels. Although a majority of users of this guide will use colour and form, including body proportions, to identify species, it is counts of specific scales that often set species apart. Figures 1 & 2 (p.6) show important scales that can be counted in order to confirm identification. Important scales include the midbody scale rows (dorsals), scales along the belly (ventrals) and those under the tail (subcaudals). A number of scales of the head are of importance in species identification, some of which are shown here.

Counts should not be made on live snakes for obvious reasons (potential danger to the investigator, stress to the snake, etc.); counts of scales can be made easily (after some practice) on both dead snakes (such as road-kills) and shed skins. Readers are encouraged to read more technical works in order to reliably identify species, as may be required in cases of bites from unknown snakes, or simply to know what species is found in a particular area.

Figure 1 Scales of a snake's dorsum and head.

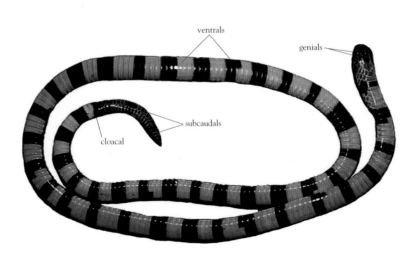

Figure 2 Scales of a snake's venter.

DEALING WITH SNAKE BITES

A number of snake species in Southeast Asia can deliver bites that are life-threatening to humans. These highly venomous snakes are found in forests and fields, in coastal waters and also out in the open sea. Deaths from snake bites are generally caused by a relatively small group of snakes, most of which are adapted to human-modified environments or live in forest edges. In this section, I list precautionary measures for avoiding snake bites and for dealing with cases of snake envenomation, as relevant to the region.

The majority of Southeast Asia's snakes are non-venomous, and the chances of getting bitten are more remote than those of being in a motor accident or drowning. It is important, nonetheless, to be able to identify the major venomous snake groups. Vipers (such as *Daboia*) and pit vipers (such as *Parias*, *Popeia*, *Trimeresurus* and *Tropidolaemus*) are relatively slow-moving snakes, with narrow necks and enlarged heads, whose fangs can be folded when not in use. Cobras (*Naja* and *Ophiophagus*) are large, heavy-bodied snakes, with the ability to raise a hood; they have short, fixed fangs. Coral snakes (*Calliophis* and *Sinomicrurus*) and kraits (*Bungarus*) are close relatives of cobras, but cannot raise their hoods. Finally, the sea snakes (including *Hydrophis* and *Laticauda*) are large, slender- or heavy-bodied snakes that are marine or at least coastal in distribution, with just a few (e.g. *Hydrophis sibauensis*) travelling far up tidal rivers. A genus of sea snake, *Laticauda* (two species in this region), comes ashore. All sea snakes have short, fixed fangs.

Many non-venomous snakes (particularly the kukri snakes and cat snakes) can inflict a painful bite, which while not immediately fatal may lead to bleeding and secondary infection. Needless to say, a bite from a large python can be dangerous, and these giant constricting snakes, especially the Reticulated Python, have been known to kill adult humans through constriction.

Below are a few dos and don'ts to follow when visiting places where venomous or unknown snakes have been sighted:

- Don't put your hand inside cracks or holes where a snake may be sheltering.
- Wear shoes that conceal the entire foot, especially in tall grass, where large vipers may be present.
- When moving in the dark, carry a reliable torch or wear a headtorch.
- To keep snakes away from human residences, ensure surroundings are free from litter, which attracts rats, and in turn, rodent-eating snakes.

Anti-venom sera used in treating snake bites are maintained at many hospitals, clinics and primary health centres in areas where such incidents are common (e.g. farms and oil-palm estates). In the event of a venomous snake bite, the patient needs to be kept calm and warm, and taken to a hospital as quickly as possible. The region around the bite should be immobilised with a stiff cloth bandage (not a tight tourniquet), in the case of bites from cobras, kraits, coral snakes and sea snakes. Some description or photograph of the snake in question will help medical staff provide appropriate treatment, as the neurotoxic venom of cobras, kraits and coral snakes acts differently from the haemotoxic venom of vipers. It is not advisable to cut or suck a bite, as these actions are likely to complicate the treatment as well as the subsequent healing process.

Finally, remember that anti-venom serum is the only proven cure for a snake bite.

About this Book

This book deals with representative species of snake the average visitor to, or resident of, Southeast Asia is likely to encounter. A number of the species are, nonetheless, rare, and are perhaps being illustrated here for the first time in a printed work of this sort. The aim of the volume is to aid rapid field identification – useful for biodiversity surveys, necessary for conservation and management, or simply enjoyable for anyone interested in fauna. The cut-off date for the checklist in this work was 31 July 2020.

For each species covered, the heading provides the following details: a common English name (the majority of which are from published sources); the current scientific name; the maximum total length attained; and, where they exist, vernacular and 'book' names (names applied by herpetologists to particular species in published works, which may not necessarily be used by indigenous people of a certain region) in some of the local languages (including Chinese languages, Dusun, Iban, Bahasa Malaysia/Brunei/Indonesia, Kelabit, Sundanese and Thai). In the species descriptions, the following information is given: colours and morphological characters used to aid field identification; distributional range within the area covered by this work, and notes on occurrence in extralimital areas; and brief notes on habits and behaviour, including habitat associations, elevational range, diet and reproduction, when known. The conservation status of each species according to the 2020 IUCN Red List of Threatened Species, Red List Categories, Version version 2020-2 (www.iucnredlist.org), is given in the Checklist of Southeast Asian Snakes (pp. 158–169).

Abbreviations

asl	above sea-level	m	metre
c.	circa	mm	millimetre
cm	centimetre	N	north
E	east	S	south
IUCN	International Union for Conservation of Nature and Natural Resources	W	west

Glossary

Anterior Toward front of body; opposite of posterior.
Arboreal Living in trees.
Autotomy Spontaneous or reflexive separation of a body part (typically, a tail).
Chevron V-shaped mark.
Cloaca Chamber into which intestinal, urinary and reproductive ducts discharge contents.
Clutch Entire compliment of eggs or neonates from a single female.
Concave Bent inwards, rounded.
Conical scales Cone-shaped scales.
Convex Bent outwards, rounded.

Cryptic Camouflaged or hidden.

Dimorphism Difference in morphology between members of the same species.

Diurnal Active during the day.

Dorsal Toward upper surface of the body; scale in this area.

Dorsal crest Ridge of highly modified (often conical) scales along back.

Dorsum Back, or dorsal surface, of body.

Fang Recurved, elongate teeth on upper jaw, through which venom passes.

Furrow Well-defined groove.

Granular scales Small, convex, non-overlapping scales, typically with a pebbly appearance.

Hood Expanded skin behind head, especially in cobras.

Imbricate With regularly arranged, overlapping edges, like tiles on a roof.

Juvenile Young or sexually immature individual.

Keel Raised ridge down back, tail or scale.

Lateral Pertaining to side of body.

Nuchal Relating to the back of the neck.

Nuchal venom gland Integumental glands in paravertebral region of neck of several species of snakes.

Ocellate With eye-like markings.

Oviparity Reproduction through production of eggs that have membranes and/or shells.

Ovoviviparity Reproduction through production of live young that hatch from eggs within female oviducts.

Paravertebral stripe Stripe on one side of midline of dorsum.

Parthenogenesis Form of asexual reproduction involving development of embryos without fertilisation.

Posterior Toward rear of body; opposite of anterior.

Prehensile Able to grasp objects.

Reticulation Colour pattern resembling mesh of a net.

Scalation Pattern of scales on body or on a specific part of body.

Scute Enlarged scale.

Serrated With a saw-like appearance.

Sexual dimorphism Condition in which males and females have distinctly different forms.

Snout-vent length Measurement between snout tip and vent.

Spinose Sharp, pointed shape like a thorn.

Sub-fossorial Habit of living under a substrate, such as in leaf litter, under fallen objects.

Temporal Scale behind post-ocular.

Total length Measurement between snout tip and tail tip.

Venom Substance capable of producing toxic reaction when introduced into tissue.

Ventrum Underside of body.

ACROCHORDIDAE – WART SNAKES
This family includes three species worldwide, two of them in the region. They are recognisable in showing heavy bodies; loose, folded skin with rough, granular scales and bristle-tipped tubercles; valvular nostrils; eyes positioned on top of the head; and a flap for closing the lingual opening of mouth – all adaptations for a highly aquatic mode of life. They inhabit fresh waters and sea coasts, and are nocturnal, secreting themselves beneath fallen logs and other debris underwater, and emerging to hunt crabs, fish and other snakes at night. Large-growing species are harvested for their durable skins and also for their flesh, while at other localities they are killed by fishermen on account of their fish diet.

Wart Snake ▪ *Acrochordus granulatus* 100cm
(Bahasa Malaysia: Ular Kadut. Bahasa Indonesia: Ular Air Tawar Kecil. Iban: Ular Paiie. Thai: Ngu Pai-ki-reu)

DESCRIPTION Top of body is olive, blue or blackish grey, and is marked with distinct transverse cream bands, especially in juveniles, that may sometimes persist in adults. Body is stout but compressed; head is indistinct from neck, and covered with small juxtaposed scales; eyes are tiny with a vertical pupil; mid-body scale is largest on vertebral region; tail is short and prehensile; a distinct fold of skin is present along middle of belly. **DISTRIBUTION** Myanmar, Thailand, Peninsular Malaysia, Sumatra, Borneo, Java. Extralimitally: from Indian sub-continent to Southeast Asia, New Guinea and Australia. **HABITS AND HABITAT** Coastal regions, such as estuaries, mangroves and sea coasts. Diet includes crabs, eels, burrowing gobies and other snakes. One population on Lake Taal, Luzon, in the Philippines lives in a freshwater lake. Ovoviviparous, producing 6–12 neonates (360–400mm).

Elephant Trunk Snake ■ *Acrochordus javanicus* 200cm

(Bahasa Malaysia/Indonesia: Ular Belalai Gajah. Hakka Chinese: Nai She. Iban: Ular Pai. Thai: Ngu Nguang-chang)

DESCRIPTION Top of body is greyish black, the head with darker lines; 2 diffuse longitudinal stripes and elongated dark blotches are present on flanks; belly is cream. Compared to its relative, the Wart Snake (see opposite), body is extremely stout and slightly compressed; head is indistinct from neck; forehead scales are small and rough; eyes are small with a vertical pupil; dorsals are keeled; mid-body scale rows are largest around vertebrals; tail is short but prehensile. **DISTRIBUTION** Thailand, Peninsular Malaysia, Singapore, Sumatra, Borneo, Java. Extralimitally: Cambodia, Vietnam. **HABITS AND HABITAT** Freshwater wetlands, including peat swamps and black-water rivers, plus ditches and canals. Diet comprises fish, including eels and catfish. Ovoviviparous, producing 6–48 neonates (290–460mm). An interesting aspect of its reproductive biology is its capacity, as documented once, to produce embryos without mating. Termed 'parthenogenesis', this phenomenon has also been reported in a few other snakes.

ANOMOCHILIDAE – GIANT BLIND SNAKES
This family of burrowing snakes is restricted to Sundaland, and is represented by three
species, all with sub-cylindrical bodies. They lack a chin groove, and lack teeth on
the pterygoid and palatine bones. Poorly known, they inhabit lowland and montane
forests, and are at least superficially similar to the Asian pipesnakes (Cylindrophiidae).

Kinabalu Giant Blind Snake ▪ *Anomochilus monticola* 52.1cm
(Bahasa Malaysia/Bahasa Indonesia: Ular Tanah)

DESCRIPTION Top of body is blue-black, and lacks pale lateral lines and large pale
blotches on either side of vertebral region; has a transverse yellow bar across snout; has a

series of isolated pale yellow scales on
flanks; belly is dark brown. Body is stout,
being rounded in cross section; head is
small and indistinct from neck; forehead
is covered with large scales; eyes are
small; tail is short and conical; dorsals
are smooth, and slightly larger than
ventrals at same level. **DISTRIBUTION**
Gunung Kinabalu, Sabah, Borneo.
HABITS AND HABITAT Sub-montane
forests (1,450m asl). Edges of water
bodies and in human-modified areas.
Diet includes arthropods. Reproductive
habits are unknown.

Burmese Pipe Snake ▪ *Cylindrophis burmanus* 32cm

DESCRIPTION Back of body black with narrow and alternating bands; narrow ring encircling nape; light postocular streak; irregular rows of paravertebral spots along length of body; undersurface of tail bright orange. **DISTRIBUTION** Myanmar (Pyinmana, Thandoung and Sahmaw, in Upper and Myitkyina District). **HABITS AND HABITAT** Inhabits lowland forest areas. Diet and reproductive habits unstudied.

Lined Pipe Snake ■ *Cylindrophis lineatus* 982mm

DESCRIPTION Top of body bright reddish-pink; series of longitudinal red or yellow stripes from back of head to base of tail; forehead has scattered dark spots. Body robust and elongate, and flattened when displaying;

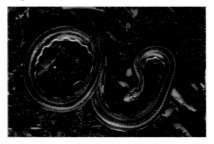

mental groove present; head long, blunt and indistinct from neck; eyes reduced; tail tapers to narrow point; dorsal scales smooth; cloacal scute divided. **DISTRIBUTION** North-eastern Borneo (Sarawak State). **HABITAT AND HABITS** Found in lowland forests and low hills at up to around 400m above sea level. Subfossorial. Diet and reproductive habits unknown.

Common Pipe Snake ■ *Cylindrophis ruffus* 90cm

(Bahasa Malaysia: Ular Kepala Dua, Ular Tanah. Bahasa Indonesia: Ular Kepala Dua. Iban: Ular Bangkit, Ular Untup. Sundanese: Oraj Teropong. Thai: Ngu Kon-kob)

DESCRIPTION Top of body is black, typically with a pale collar; cream bands on dorsum are present in some populations; belly has black cross-bars. Body is robust, elongate and flattened when displaying; mental groove is present; head is short, blunt and indistinct from neck; eyes are reduced with a rounded or vertical pupil; tail tapers to a narrow point; dorsals are smooth; cloacal scute is divided. **DISTRIBUTION** Myanmar, Thailand, Peninsular Malaysia, Sumatra, Pulau Bangka, Pulau Belitung, Riau Archipelago, Borneo, Java. Extralimitally: Laos, Cambodia, Vietnam, China, Sangihe and Sula archipelagos, Sulawesi. **HABITS AND HABITAT** Low, swampy areas within forested habitats, salt-water lagoons and agricultural fields (<1,676m asl). Sub-fossorial. Diet comprises snakes and eels. Ovoviviparous, producing 5–13 neonates (205mm).

PYTHONIDAE – PYTHONS
This family includes the largest snakes in the region, one species reaching 10m in total length. Famous in literature and legend, the pythons are swallowers of small and mid-sized mammalian prey, with at least two species reaching sizes large enough to subdue and swallow adult humans, although such instances are rare. Pythons show teeth on their pre-maxilla, and also have supra-orbital bones on the dorsal margin of orbit. In addition, they have rows of heat-sensing labial organs and vestigial pelvic and hind limb bones (visible as paired spurs on each side of the cloaca). All snakes in this family are egg-layers. Their global distribution covers the Old World tropics and subtropics, from Africa, through Asia and the archipelagos of Australasia, to Australia.

Reticulated Python ▪ *Malayopython reticulatus* >1,000cm

(Mandarin Chinese: Mang Seh. Hakka Chinese: Kim Seh. Bahasa Brunei: Ular Penalan. Bahasa Malaysia: Ular Sawa Batik, Ular Sawa Cindai. Bahasa Indonesia: Ular Sanca Batik, Ular Saab. Iban: Ular Sawah. Thai: Ngu Leuam)

DESCRIPTION Top of body is yellow or brown, with dark, rhomboidal markings; a black median line runs from snout to nape; an oblique line runs from posterior of eye to corner of mouth; belly is yellow with small brown spots. Body is relatively elongated and slender, except in large individuals; head is distinct from neck; some infralabials are equipped with pits; eyes are small with a vertical pupil; cloacal spurs are present in both sexes.

DISTRIBUTION Myanmar; Thailand; Peninsular Malaysia; Singapore; Sumatra; Pulau Bangka; Pulau Belitung; Pulau Weh; Pulau Enggano; Pulau Nias; Mentawai, Natuna and Riau archipelagos; Borneo; Java; Bali. Extralimitally: Laos, Cambodia, Vietnam, Nicobar Islands, Ambon, Anambas Archipelago, Babi, Batjan, Banda Besar, Bankak, Boano, Buru, Butung, Flores, Halmahera, Haruku, Lang, Lombok, Obira, Saparua, Seram, Sula Archipelago, Sulawesi, Sumba, Sumbawa, Tanimbar, Ternate, Timor and Verlate, Basilan, Bohol, Calamian Islands, Cebu, Leyte, Luzon, Mindanao, Mindoro, Negros, Palawan, Panay, Polillo, Samar, Tawi-Tawi, Sulu Archipelago. **HABITS AND HABITAT** Its natural

habitat is forests, especially at the water's edge, where it waits in ambush for deer and pigs; may now also be common in cities and towns, where it inhabits sewers. Mostly found on the ground or in water, but can climb trees, and mostly active at night. Diet comprises warm-blooded animals, such mammals and birds, although lizards may also be eaten; occasional reports of attacks on humans. Oviparous, producing clutches of 14–124 eggs (90–93 × 58–62mm).

Southeast Asian Rock Python ■ *Python bivittatus* 574cm
(Bahasa Malaysia: Ular Sawa. Bahasa Indonesia: Ular Sanca Kembang. Thai: Ngu Lam)

DESCRIPTION Top of body is dark brown or yellowish grey, with a series of 30–40 large, irregular squarish, black-edged, dark chocolate-grey patches on top and sides of body; has dark and dark grey dorsal and lateral spots; has a sub-ocular stripe; belly is grey with dark spots on outer scale rows. Body is thick and cylindrical; head is lance-shaped, and distinct from neck; sensory pits are present in rostrals and on some supralabials and infralabials; spurs are small; tail is short and prehensile; cloacal spurs are present. **DISTRIBUTION**

Myanmar, Thailand, Java, possibly N Peninsular Malaysia. Extralimitally: Laos, Cambodia, Vietnam, E India, Bangladesh, Nepal, S China, Sulawesi, Sumbawa. Introduced to SE USA. **HABITS AND HABITAT** Dry seasonal forests; sometimes found in towns and villages. Eats warm-blooded prey, such as monkeys, goats and calves. Oviparous, producing clutches of 30–58 eggs (120 × 60mm).

Bornean Short Python ■ *Python breitensteini* >200cm
(Bahasa Malaysia/Indonesia: Ular Sawa Darah. Iban: Ular Ripong)

DESCRIPTION Top of body is pale yellow or tan, with dark sub-rectangular blotches about as wide as body, becoming darker towards tail, or with a fully dark top that turns black posteriorly; scattered pale spots are present on vertebral region, and are more numerous at back of body, where they are elongate and form a vertebral stripe; has a black stripe evident between internasals and occipital, fusing with dark pattern on neck; sides of head are darker than forehead, with dark flecks and a broad dark post-ocular stripe; a pale post-ocular stripe

runs to angle of jaws; chin and belly are plain cream, sometimes with brown spots. Body is short and robust; head is elongate, flattened and distinct from neck; eyes are small with a vertical pupil; vertebral region is ridged; some infralabials have weak pits; tail is short; cloacal spurs are present. **DISTRIBUTION** Borneo. **HABITS AND HABITAT** Lowland rainforests, peat swamps and heath forests, up to sub-montane limits (<1,000m asl). Edges of rivers, swamps and marshes. Diet includes small mammals and birds. Oviparous, clutches comprising 12 eggs (size unknown).

Brongersma's Short Python ■ *Python brongersmai* 260cm
(Thai: Ngu Lam Pak Ped)

DESCRIPTION Top of body is red, reddish brown, charcoal-grey, pale grey or brown; an indistinct narrow, dark stripe is present along middle of forehead; supralabials are dark; a pale, narrow post-ocular stripe extends to angle of jaws; dorsal pattern comprises vertebral spots; dark blotches on flanks are rounded and set within paler areas; belly is anteriorly cream and posteriorly dark, with grey smudges and blotches. Body is short and robust; head is elongate, flat and distinct from neck; vertebral region is ridged; tail is short; cloacal spurs are present. **DISTRIBUTION** Thailand, Peninsular Malaysia, Singapore, Sumatra, Pulau Bangka, Mentawai Archipelago. Extralimitally: Laos, Cambodia, Vietnam. **HABITS AND HABITAT** Lowland forests, at edges of streams (<1,330m asl). Diet includes small mammals and birds. Oviparous, laying clutches of 10–15 eggs (size unknown).

Sumatran Short Python ■ *Python curtus* >200cm
(Bahasa Indonesia: Ular Sawa)

DESCRIPTION Top of body is brownish grey, with a series of longitudinal dark, irregular sub-rectangular blotches, sometimes as broad as body; flanks have a longitudinal series of large blotches with black edges; sides of snout have dark stripes; post-ocular pattern comprises black triangular blotches, widening to 4–6 scales at angle of jaws and coalescing with anterior labial blotch of neck; has a pale post-ocular stripe with dark smudges; chin and belly are plain cream or white. Body is short and robust; head is elongate, flat and distinct from neck; vertebral region is ridged; tail is short; cloacal spurs are present. **DISTRIBUTION** Sumatra. **HABITS AND HABITAT** Lowland forests, up to sub-montane limits (c. 1,800m asl). Streams and forest floor. Diet comprises small mammals and possibly also birds. Oviparous, clutches comprising 10–12 eggs (size unknown).

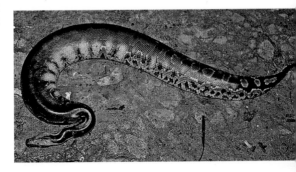

Xᴇɴᴏᴘᴇʟᴛɪᴅᴀᴇ – Sᴜɴʙᴇᴀᴍ Sɴᴀᴋᴇs

Two living representatives of this family are known, both with highly iridescent dorsals, depressed snout, sub-cylindrical body, short tail, large scales on forehead and reduced ventrals. They are found in lowland forests and are sub-fossorial, living under loose leaf litter or fallen objects on the forest floor, and they feed on small vertebrates that they kill by constriction. They are known from Southeast Asia and eastern China.

Sunbeam Snake ▪ *Xenopeltis unicolor* 114cm
(Bahasa Malaysia: Ular Pelangi. Thai: Ngu Saeng-a-tit)

DESCRIPTION Top of body is iridescent brown, each scale light-edged; juveniles have a pale collar; belly is white or cream. Body is robust and cylindrical; head is slightly distinct from neck; snout is rounded and depressed; eyes are small with a vertical pupil; tail is short; dorsals are smooth; cloacal scute is divided. **DISTRIBUTION** Myanmar, Thailand, Peninsular Malaysia, Singapore, Sumatra, Pulau Bangka, Pulau Belitung, Riau Archipelago, Pulau Natuna Besar, Pulau Sipura, Pulau Simeulue, Pulau Siberut, Borneo, Java. Extralimitally: Laos, Cambodia, Vietnam, China, Nicobar Archipelago, Palawan, Sulu Archipelago. **HABITS AND HABITAT** Lowland dipterocarp forests, reaching sub-montane limits (<1,402m asl). Terrestrial and sub-fossorial. Diet comprises rodents, birds, lizards and frogs. Oviparous, clutches including 3–17 eggs (18 × 58mm).

COLUBRIDAE – 'TYPICAL' SNAKES
The majority of snakes in Southeast Asia belong to this family, which until recently also included the water snakes, slug-eating snakes and several other groups, now reassigned. They can be told apart by their large forehead scales, solid (not grooved) maxillary teeth, laterally placed nostrils and well-developed ventrals. They may be either oviparous or ovoviviparous, and are also found beyond Southeast Asia, in temperate, subtropical and tropical parts of the world.

Speckle-headed Vine Snake ■ *Ahaetulla fasciolata* 169cm
(Iban: Ular Bunga Merisian. Thai: Ngu Kieo Hua Lai Kra)

DESCRIPTION Top of body is light brown, grey or pinkish tan, with numerous narrow, oblique, dark bands on anterior of body; forehead has elongated or curved dark markings; belly is dark grey. Body is slender; snout is long, ending in curled rostral; eyes are large with a horizontal pupil; tail is long, with a prehensile tip; vertebrals are enlarged; dorsals are smooth.
DISTRIBUTION Thailand, Peninsular Malaysia, Singapore, Sumatra, Natuna and Riau archipelagos, Borneo. **HABITS AND HABITAT** Forested and semi-urban habitats (<900m asl). Arboreal, in thick undergrowth and other low vegetation. Diet includes lizards and frogs. Ovoviviparous (numbers and size of neonates unknown).

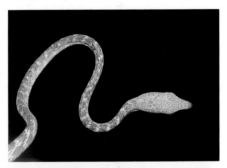

River Vine Snake ■ *Ahaetulla fronticincta* 98cm

DESCRIPTION Top of body is either bright green or brownish yellow, with black and white interstitial skin, forming oblique lines; belly is pale green or olive; has a white streak along lower flanks; forehead is with or without black spots. Body is slender; snout is long; eyes are large with a horizontal pupil; tail is long, with a prehensile tip; dorsals are smooth. **DISTRIBUTION** Myanmar. **HABITS AND HABITAT** Edges of creek and river mouths, near mangroves. Arboreal, in thick undergrowth and other low vegetation. Diet comprises surface-feeding fish. Ovoviviparous, producing clutches of 7 neonates (size unknown).

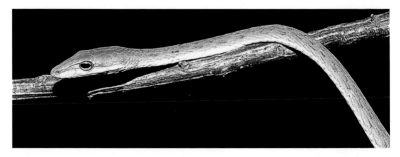

Malayan Vine Snake ■ *Ahaetulla mycterizans* 92cm
(Bahasa Malaysia: Ular Cemeti Hijau. Thai: Ngu Kieo Hua Ching-chok)

DESCRIPTION Top of body is bright green, greyish green or brown; in green morph, belly is white with paired longitudinal green lines and, sometimes, a green line along middle. Body is slender; snout is elongated; groove is present along snout; eyes are large with a horizontal pupil; tail is long, with a prehensile tip; dorsals are smooth. **DISTRIBUTION** Thailand, Peninsular Malaysia, Sumatra, Java. **HABITS AND HABITAT** Lowland forests (<500m asl). Arboreal, in thick undergrowth and other low vegetation. Diet comprises lizards and birds. Reproductive habits are unstudied.

Long-nosed Vine Snake ■ *Ahaetulla nasuta* 200cm
(Thai: Ngu Kieo Pak Nab)

DESCRIPTION Top of body is bright green or, less often, olive-brown, with a longitudinal yellowish line along outer margin of ventrals; belly is pale green; iris is yellow. Body is elongate and slender; snout is long, with a rostral appendage; a groove is present in front of eyes; supra-ocular is divided horizontally; eyes are large with a horizontal pupil; tail is long, with a prehensile tip; dorsals are smooth; cloacal scute is divided. **DISTRIBUTION** Myanmar, Thailand. Extralimitally: India, Bangladesh, Bhutan, Nepal, Sri Lanka, Laos, Cambodia, Vietnam. **HABITS AND HABITAT** Lightly forested habitats, including gardens, frequenting trees and bushes (<1,800m asl). Arboreal, in thick undergrowth and other low vegetation. Diet comprises tadpoles, lizards, birds and small mammals. Ovoviviparous, producing 3–23 neonates (200–440mm).

Oriental Vine Snake ▪ *Ahaetulla prasina* 197cm

(Bahasa Malaysia: Ular Pucuk. Bahasa Brunei: Ular Daun, Ular Kunyet. Bahasa Indonesia: Ular Pucuk. Iban: Ular Bungai, Puchok Pisang. Kelabit: Selangoi Bata. Sundanese: Oraj Gadung. Thai: Ngu Kieo Hua Ching-chok Pa)

DESCRIPTION Top of body is usually green, but may also be brown, yellow, dark grey or golden yellow, speckled with black; a yellow stripe runs along lower flanks; belly is light green or dark grey. Body is slender; snout is elongated, with a groove running along it; eyes are large

with a horizontal pupil; tail is long, with a prehensile tip; dorsals are smooth; cloacal scute is divided. **DISTRIBUTION** Myanmar; Thailand; Peninsular Malaysia; Singapore; Sumatra; Mentawai, Riau and Natuna archipelagos; Pulau Bangka; Pulau Belitung; Pulau Sibutu; Borneo; Java; Bali. Extralimitally: Bhutan, E India, Bangladesh, Laos, Cambodia, Vietnam, China, Sulu Archipelago and other islands of the Philippines. **HABITS AND HABITAT** Edges of forests and gardens (<2,100m asl). Arboreal, in thick undergrowth and other low vegetation. Diet comprises lizards and birds. Ovoviviparous, producing 4–10 neonates (240–490mm).

Dice-like Rat Snake ▪ *Archelaphe bella* 80cm

DESCRIPTION Top of body is brown or greyish brown, with brown saddle-shaped blotches or with transverse or oblique cross-bars; forehead is pale yellow to yellowish brown, with

black Y-shaped mark or lighter streak edged with black; labials are dark-edged; belly is yellow with irregular large black blotches on each ventral. Body is slender; head is indistinct from neck; snout is rounded; eyes are small with a vertical pupil; dorsals are smooth; tail is short; cloacal scute is divided. **DISTRIBUTION** Myanmar. Extralimitally: Vietnam (reports from NE India are as yet unverified). **HABITS AND HABITAT** Sub-montane and montane forests (1,500–2,000m asl). Diet and reproductive biology are unstudied.

Iridescent Snake ■ *Blythia reticulata* 51.4cm

DESCRIPTION Top of body is olive to dark, and highly iridescent; scales are sometimes light-speckled or light-bordered; juveniles show a yellowish-cream collar and a gap on the

dark vertebral line; ventrals are grey with a pale posterior edge. Body is slender and elongate; head is barely distinct from neck; eyes are moderate in size with a rounded or vertical pupil; dorsals are smooth and glossy, and lack apical pits; tail is short, with an acute tip; cloacal scute is divided. **DISTRIBUTION** Myanmar. Extralimitally: E India. **HABITS AND HABITAT** Wet evergreen forests (<1,040m asl). Sub-fossorial, under fallen logs and litter. Diet includes earthworms and, probably, soil arthropods. Oviparous, clutches comprising 6 eggs (size unknown).

Green Cat Snake ■ *Boiga cyanea* 187cm
(Thai: Ngu Kieo Bon)

DESCRIPTION Top of body is emerald-green in adults, and reddish brown or olive with a green forehead in juveniles; interstitial skin is black; gular region is sky-blue; belly

is greenish white or greenish yellow, plain or spotted with dark green; iris is brownish grey. Body is slender, elongate and laterally compressed; head is large and distinct from neck; eyes are large with a vertical pupil; vertebral region has a low ridge; vertebral scale rows are enlarged; scales are smooth, with apical pits; cloacal scute is entire. **DISTRIBUTION** Myanmar, Thailand, Peninsular Malaysia. Extralimitally: Bangladesh, Bhutan, Nepal, E India, Nicobar Islands, S China, Laos, Cambodia, Vietnam. **HABITS AND HABITAT** Forests and disturbed habitats (150–2,100m asl). Arboreal, on trees and in undergrowth. Diet comprises frogs, birds and their eggs, lizards, snakes and small mammals. Oviparous, laying clutches of 4–10 eggs (40–48 × 15–21mm).

Dog-toothed Cat Snake ■ *Boiga cynodon* 280cm
(Bahasa Malaysia: Ular Telor. Bahasa Indonesia: Ular Kucing Bergigi Panjang. Iban: Ular Blidah, Ular Kengkang Mas. Thai: Ngu Sae Hang-ma)

DESCRIPTION Top of body is brownish tan or yellowish brown, with dark brown or reddish-brown bands that darken posteriorly; juveniles are paler than adults; a dark post-ocular stripe is present. Body is slender, elongate and laterally compressed; head is distinct from neck; snout is short and rounded; eyes are large with a vertical pupil; vertebrals are distinctly enlarged; dorsals are smooth; cloacal scute is entire. **DISTRIBUTION** Thailand, Peninsular Malaysia, Singapore, Sumatra, Pulau Nias, Pulau Belitung, Pulau Bangka, Mentawai Archipelago, Borneo, Java, Bali. Extralimitally: Lesser Sunda Islands, Philippines. **HABITS AND HABITAT** Lowland forests and forest edges, occasionally human settlements. Arboreal, on trees and in dense undergrowth. Diet includes lizards, birds and their eggs, and small mammals. Oviparous, laying 6–23 eggs (size unknown).

Mangrove Cat Snake ■ *Boiga dendrophila* 250cm
(Bahasa Malaysia/Indonesia: Ular Bakau, Ular Taliwangsa. Iban: Ular Bangkit, Ular Chinchin Mas. Sundanese: Oraj Taliwangsa. Thai: Ngu Plong-tong)

DESCRIPTION Top of body is black, with 35–45 narrow yellow transverse rings on body and 10 on tail; labials and gular region are yellow; belly is grey. Body is large, robust and compressed; head is distinct from neck; snout is short and rounded; eyes are large with

a vertical pupil; dorsals are smooth; cloacal scute is entire. **DISTRIBUTION** Thailand, Peninsular Malaysia, Singapore, Sumatra, Pulau Belitung and Batu Archipelago, Pulau Nias, Borneo, Java, Bali. Extralimitally: Cambodia, Vietnam, islands of the Philippines and central Indonesia. **HABITS AND HABITAT** Mangroves and peat swamps, lowland mixed dipterocarp forests and edges of human settlements. Arboreal, on trees and in dense undergrowth. Diet comprises birds, and their eggs and nestlings, as well as frogs, lizards, other snakes, mouse deer and tree shrews. Oviparous, laying 4–15 eggs (45.5–51 × 24.5–25mm).

White-spotted Cat Snake ■ *Boiga drapiezii* 210cm
(Thai: Ngu Dong-ka Tong)

DESCRIPTION Top of body is variable, ranging from olive-grey to reddish brown; vertebral region is marked with paired pink spots anteriorly, these sometimes fused to form a line; pink or cream spots are also present on flanks; forehead has dark speckling. Body is relatively long and rather slender; head is distinct from neck; eyes are large with a vertical pupil; vertebrals are enlarged; dorsals are smooth; cloacal scute is entire.

DISTRIBUTION Thailand, Peninsular Malaysia, Singapore, Sumatra, Pulau Bangka, Mentawai Archipelago, Borneo. Extralimitally: Vietnam, central islands of Indonesia and the Philippines.

HABITS AND HABITAT Lowland forests and mid-hills (<1,000m asl). Arboreal, on lower strata of trees and shrubs. Diet includes birds and their eggs, frogs, lizards and large insects. Oviparous, laying eggs (number and size unknown) in termite-infested wood.

Jasper Cat Snake ■ *Boiga jaspidea* 150cm
(Iban: Ular Banjang. Thai: Ngu Kra)

DESCRIPTION Top of body is brown, reddish brown or grey-brown, with paired rows of dark spots or bars on flanks and a greyish-red vertebral stripe. Body is slender and laterally compressed; head is large, distinct from neck; eyes are large with a vertical pupil; vertebrals are enlarged; dorsals are smooth; cloacal scute is enlarged.

DISTRIBUTION Thailand, Peninsular Malaysia, Singapore, Sumatra, Pulau Nias, Mentawai Archipelago, Pulau Bangka, Borneo, Java. Extralimitally: Vietnam.

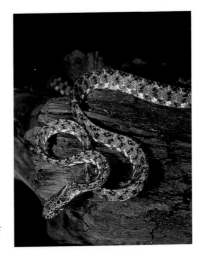

HABITS AND HABITAT Lowland forests and peat swamps (<1,524m asl). Arboreal, on trees and in dense undergrowth. Diet includes lizards, small mammals, birds and their eggs, and other snakes. Oviparous, laying 6 eggs (38–39 × 18–19mm) in nests of tree-dwelling termites.

Many-spotted Cat Snake ■ *Boiga multomaculata* 187cm
(Bahasa Indonesia: Ular Kucing Loreng. Thai: Ngu Me-ta-ngao Rang-nok)

DESCRIPTION Top of body is grey-brown; 2 black-edged brown lines run from snout to back of head; a series of irregular brown blotches is present on dorsum, plus smaller brown marks on flanks; belly is greyish brown with small brown spots. Body is slender and laterally

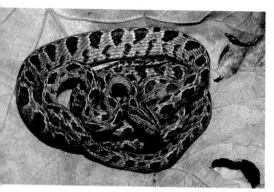

compressed; head is large and distinct from neck; eyes are large with a vertical pupil; dorsals are smooth; cloacal scute is entire. **DISTRIBUTION** Myanmar, Thailand. Extralimitally: E India, Bangladesh, E China, Laos, Cambodia, Vietnam. **HABITS AND HABITAT** Lowlands and sub-montane forests (<1,500m asl). Arboreal, in short trees, bushes and bamboo groves. Diet includes birds and lizards. Oviparous, laying clutches of 4–8 eggs (26–32 × 11–12mm).

Black-headed Cat Snake ■ *Boiga nigriceps* 200cm
(Iban: Ular Banjang. Thai: Ngu Tong-fi)

DESCRIPTION Top of body is straw-brown, olive-brown or reddish brown; forehead is often darker, as is tail; labials are cream or yellow; belly is cream, darkening posteriorly.

Body is robust and laterally compressed; head is large and distinct from neck; eyes are large with a vertical pupil; dorsals are smooth; cloacal scute is entire. **DISTRIBUTION** Thailand, Peninsular Malaysia, Sumatra, Pulau Nias, Pulau Simeulue, Borneo, Java, Mentawai Archipelago. **HABITS AND HABITAT** Lowland forests (<800m asl). Arboreal, on trees and in undergrowth. Diet includes birds and other snakes. Oviparous, producing 3 eggs (48 × 17mm).

Tawny Cat Snake ■ *Boiga ochracea* 110cm

DESCRIPTION Top of body is reddish brown, ochre or coral-red, plain or with a poorly defined dark transverse line; a dark post-ocular streak extends to angle of jaws; labials are yellow or cream; belly is yellow anteriorly, light grey posteriorly. Body is slender; head is large and distinct from neck; eyes are large with a vertical pupil; dorsals are smooth with apical pits; vertebral scale row is greatly enlarged; subcaudals are paired; cloacal scute is entire.
DISTRIBUTION Myanmar, Thailand. Extralimitally: Bangladesh, Bhutan, E India, Nepal. **HABITS AND HABITAT** Forested mid-hills and submontane limits, as well as parks and gardens (350–1,400m asl). Arboreal, on bushes and in other undergrowth. Diet comprises birds and their eggs, mammals and lizards. Oviparous (number and size of eggs unknown).

Assamese Cat Snake ■ *Boiga quincunciata* 155cm

DESCRIPTION Top of body is yellow or greyish brown, finely speckled with dark brown; a vertebral series of dark brown or black spots or blotches is present, each marking 5–8 scales wide, and the scales are edged with white; flanks are speckled or spotted with brown; nape has 3 longitudinal stripes; forehead is brown, with white-edged black frontals and parietals; a black post-ocular stripe runs to angle of jaws; belly is yellowish white, intensively speckled with brown. Body is slender and elongate; head is distinct from neck; eyes are large with a vertical pupil; dorsals are smooth; vertebrals are enlarged.
DISTRIBUTION Myanmar. Extralimitally: E India.
HABITS AND HABITAT Wet evergreen forests. Arboreal, on undergrowth, especially bamboo. Diet and reproductive habits are unstudied.

Thai Cat Snake ■ *Boiga siamensis* 170cm
(Thai: Ngu Sae Hang-ma Tao)

DESCRIPTION Top of body is light brown, with 87–98 V-shaped dark brown bands, more distinct anteriorly; posteriorly, bands are bar-shaped with little posterior extension; scales are flecked with white, those comprising dark bands light at posterior tips; flanks have alternating dark and light spots; belly is light brown; forehead is mid-brown; dark streak runs from posterior margin of eye to beyond last supralabial; 2 black stripes extend on either side of vertebral row to 1st dark band and are continuous with it; belly is yellowish brown or greyish brown. Body is slender, elongated and laterally compressed; head is distinct

from neck; eyes are large with a vertical pupil; dorsals are smooth; cloacal scute is entire. **DISTRIBUTION** Myanmar, Thailand. Extralimitally: E India, Bangladesh, Laos, Cambodia, Vietnam. **HABITS AND HABITAT** Lowland and sub-montane evergreen forests (<1,780m asl). Arboreal, on trees as high as c. 6m. Diet comprises small rodents, and birds and their eggs. Reproductive habits are unstudied.

Bicoloured Reed Snake ■ *Calamaria bicolor* 45cm

DESCRIPTION Top of body is blue-black or dark brown, unpatterned or with dark cross-bands; forehead is dark brown, sometimes with 2 oblique dark bands crossing yellow labials; belly is typically plain yellow or spotted with black. Body is slender and cylindrical; head is short and indistinct from neck; nasals point forward; eyes are small with a rounded pupil; tail is thick, tapering from base; dorsals are smooth; cloacal scute is entire. **DISTRIBUTION** Borneo, Java. **HABITS AND HABITAT** Forests in mid-hills (<1,200m asl). Terrestrial. Diet and reproductive habits are unstudied.

Bornean Reed Snake ■ *Calamaria borneensis* 37.4cm

DESCRIPTION Top of body is greyish brown, each scale pale with dark reticulation; scattered dark spots or stripes on scales on mid-dorsum; 1st scale row is yellow, sometimes with dark cross-bars; head is greyish brown with indistinct dark spots; has 1–3 yellow caudal rings; belly has a dark stripe along ventral edges, or a chequered pattern of yellow and black. Body is slender and cylindrical; head is short and indistinct from neck; eyes are small with a rounded pupil; tail is short; dorsals are smooth; cloacal scute is entire. **DISTRIBUTION** Borneo. **HABITS AND HABITAT** Lowlands and sub-montane regions. Diet and reproductive biology are unstudied.

Grabowsky's Reed Snake ▪ *Calamaria grabowskyi* 47cm

DESCRIPTION Top of body is dark brown, each scale with a darker network; scattered dark brown or yellow spots are present on back; lateral bands are composed of elongated dark spots; labials are yellow; belly is plain yellow or with varying amounts of dark

pigmentation; subcaudals are yellow with a dark median band. Body is slender and cylindrical; head is short and slightly distinct from neck; eyes are small with a rounded pupil; tail is long, tapering to a blunt tip; dorsals are smooth; cloacal scute is entire. **DISTRIBUTION** Borneo. **HABITS AND HABITAT** Sub-montane forests (1,000–1,400m asl). In leaf litter. Diet and reproductive biology are unstudied.

Lined Reed Snake ▪ *Calamaria griswoldi* 49cm

DESCRIPTION Top of body is dark brown, with blackish-brown and yellow stripes; forehead is dark brown; lower portions of supralabials are yellow; an oblique pale bar

extends from parietals to gular region; ventrals are plain yellow; subcaudals are yellow with an indistinct medial zigzag mark. Body is slender and cylindrical; head is short and indistinct from neck; eyes are small with a rounded pupil; tail is short, tapering to a sharp point; dorsals are smooth; cloacal scute is entire. **DISTRIBUTION** Borneo. **HABITS AND HABITAT** Sub-montane forests (1,200–1,800m asl). In leaf litter. Diet and reproductive habits are unstudied.

Linnaeus's Reed Snake ■ *Calamaria linnaei* 40cm

DESCRIPTION Back of body black to pale brown; dorsal scales light, with a dark network; isolated small scales on back of some individuals, spots joining to form black stripes; narrow, light chevron behind forehead; belly cream coloured with dark pigments; dorsals are smooth; cloacal scute is entire. **DISTRIBUTION** Java and Pulau Bangka. **HABITS AND HABITAT** Lowland dipterocarp forests, from mid-hills up to submontane limits, at elevation of ca. 1,500m asl. Diet unstudied. Clutches comprise 2–4 eggs, measuring 20–26 x 7–9mm. Incubation period 64–84 days.

Low's Reed Snake ■ *Calamaria lovii* 32cm

DESCRIPTION Top of body is dark brown with yellow spots or, sometimes, narrow, light stripes; has a complete yellow ring around vent; forehead is dark brown with indistinct light markings; belly is cream anteriorly; posterior ventrals are dark brown or yellow with irregular squarish black blotches; subcaudals are paired. Body is slender and cylindrical; head is short and indistinct from neck; eyes are small with a rounded pupil; tail is short and thick, tapering abruptly; dorsals are smooth; cloacal scute is entire. **DISTRIBUTION** Thailand, Peninsular Malaysia, Borneo, Java. Extralimitally: Vietnam. **HABITS AND HABITAT** Forested lowlands and mid-hills (<c. 750m asl). Terrestrial, in leaf litter. Diet and reproductive habits are unstudied

Variable Reed Snake ■ *Calamaria lumbricoidea* 64cm
(Thai: Ngu Pong-oa Lak Lai)

DESCRIPTION Top of body is black with narrow cream or yellow rings; forehead is red or pink in juveniles, turning dark or even black in older snakes; belly is yellow with black ventral scales that form bands. Body is moderately robust and cylindrical; head is short, indistinct from neck; eye small with a rounded pupil; tail is short and thick, tapering

abruptly to a narrow point; dorsals are smooth; cloacal scute is entire. **DISTRIBUTION** Thailand, Peninsular Malaysia, Singapore, Sumatra, Pulau Nias, Mentawai Archipelago, Borneo, Java. Extralimitally: Mindanao, Basilan, Leyte. **HABITS AND HABITAT** Lowland and sub-montane forests and gardens (<1,676m asl). Terrestrial, in leaf litter. Diet comprises earthworms and insect larvae. Reproductive habits are unstudied.

Brown Reed Snake ■ *Calamaria pavimentata* 49cm

DESCRIPTION Back of body brown, with narrow, dark, longitudinal stripes; black collar; scales on belly with dark lateral tips; dorsals are smooth; cloacal scute is entire. **DISTRIBUTION** Myanmar (Chin and Kachin States), Thailand, Laos, Vietnam and Peninsular Malaysia. Extralimitally: north-eastern India, southern China and Japan. **HABITS AND HABITAT** Inhabits hill forests. Diet and reproductive habits unstudied.

Red-headed Reed Snake ■ *Calamaria schlegeli* 45cm
(Bahasa Indonesia: Ular Cuvier. Thai: Ngu Pong-oa Hua Daeng)

DESCRIPTION Top of body is dark brown or black; belly is plain yellow, with a red or orange (ssp. *schlegeli*) or dark brown (ssp. *cuvier*) forehead. Body is slender and cylindrical;

head is short and indistinct from neck; nasal is as large as eye; eyes are small with a rounded pupil; tail is long and tapering; dorsals are smooth; cloacal scute is entire. **DISTRIBUTION** Thailand, Peninsular Malaysia, Singapore, Sumatra, Borneo, Java, Bali. **HABITS AND HABITAT** Lowland forests. In leaf litter. Diet comprises frogs and slugs. Oviparous (number and size of eggs unknown).

Schmidt's Reed Snake ■ *Calamaria schmidti* 28cm

DESCRIPTION Top of body is plain blackish grey, with green and blue iridescence, scales pale-margined; belly is light grey or yellow, darkening posteriorly to purple. Body is slender

and cylindrical; head is short and indistinct from neck; eyes are small with a rounded pupil; tail is short with a blunt point; dorsals are smooth; cloacal scute is entire. **DISTRIBUTION** Borneo. **HABITS AND HABITAT** Montane forests (1,370–1,570m asl). Active near streams. Diet includes earthworms. Reproductive habits are unstudied.

Yellow-bellied Reed Snake ■ *Calamaria suluensis* 30cm

DESCRIPTION Back of body brown, each scale with a dark network; some scattered scales show a dark-centred spot; forehead dark brown, with scattered darker spots; belly yellow with dark spots; undersurface of tail yellow, and may have a dark median line; dorsals are smooth; cloacal scute is entire. **DISTRIBUTION** Northern Borneo (Gunung

Kinabalu and Tawau Bills Park, Sabah). Extralimitally: islands of the Sulu Archipelago (the Philippines). **HABITS AND HABITAT** Hill dipterocarp and submontane forests, at elevations between 915–1,430m asl. Diet and reproductive habits unstudied.

Short-tailed Reed Snake ■ *Calamaria virgulata* 37cm

DESCRIPTION Top of body is dark brown; dorsal scales have a light network, with or without dark longitudinal stripes; a yellow nuchal collar is sometimes present; forehead is dark brown; supralabials are yellow; belly is cream or brownish black, with dark pigments on lateral edges of ventrals; subcaudals have dark edges and a dark median stripe. Body is slender and cylindrical; head is short and indistinct from neck; eyes are small with a rounded pupil; tail is thick and tapering; dorsals are smooth; cloacal scute is entire. **DISTRIBUTION** Sumatra, Borneo, Java. Extralimitally: Sulu Archipelago, Mindanao, Palawan, Sulawesi. **HABITS AND HABITAT** Sub-montane forests. In leaf litter. Diet is unknown. Oviparous, producing 3 eggs (26–30 × 8–8.5mm).

Ornate Flying Snake ■ *Chrysopelea ornata* 140cm
(Bahasa Malaysia: Ular Pokok Emas. Bahasa Indonesia: Ular Petola, Ular Jelotong. Thai: Ngu Kieo Lai Dok-mak)

DESCRIPTION Top of body is greenish yellow or pale green; forehead is black, with yellow and black cross-bars; scales have a dark streak, forming longitudinal black stripes; belly is pale green, marked with a series of black spots on each side. Body is slender; head is depressed and distinct from neck; eyes are large with a rounded pupil; tail is long and slender; ventrals show pronounced lateral keels; dorsals are either smooth or feebly keeled, and bear apical pits; cloacal scute is divided. **DISTRIBUTION** Myanmar, Thailand, Peninsular Malaysia. Extralimitally: India, Bangladesh, Nepal, Sri Lanka, Laos, Cambodia, Vietnam. **HABITS AND HABITAT** Lowland forests and secondary vegetation; often enters houses. Arboreal, making long glides between trees. Diet comprises lizards, bats, rodents, birds and snakes. Oviparous, producing 6–20 eggs (26–38 × 13–18mm).

Garden Flying Snake ■ *Chrysopelea paradisi* 150cm
(Bahasa Malaysia: Ular Petola. Bahasa Indonesia: Ular Pohon Paradise. Thai: Ngu Kieo Ron)

DESCRIPTION Top of body is black, centre of each scale with a green spot; vertebral region of some individuals has a row of 3–4 pink or red spots; forehead has yellow bands; belly is green with black edges. Body is slender; head is depressed and distinct from neck; eyes are large with a rounded pupil; tail is long and slender; ventrals have pronounced lateral keels; dorsals are smooth or weakly keeled; cloacal scute is divided. **DISTRIBUTION** Myanmar, Thailand, Peninsular Malaysia, Singapore, Sumatra, Mentawai Archipelago, Borneo, Java. Extralimitally: Andaman Archipelago, Sulawesi, islands of the Philippines. **HABITS AND HABITAT** Lowland to sub-montane forests (<1,524m asl). Arboreal; can make extended leaps. Diet includes lizards, and perhaps bats and small birds. Oviparous, producing 5–8 eggs (size unknown).

Twin-barred Flying Snake ■ *Chrysopelea pelias* 74cm
(Bahasa Malaysia: Ular Pokok Belang. Thai: Ngu Dog-mak Daeng)

DESCRIPTION Top of body is red or orange with yellow or cream cross-bars, these edged with black bars; forehead has 3 red cross-bars; belly is pale. Body is slender; head is depressed and distinct from neck; eyes are large with a rounded pupil; tail is long and slender; ventrals have pronounced lateral keels; dorsals are smooth or weakly keeled, and have apical pits; cloacal scute

is divided. **DISTRIBUTION** Myanmar; Thailand; Peninsular Malaysia; Singapore; Sumatra; Pulau Bangka; Mentawai, Natuna and Riau archipelagos; Borneo; Java. **HABITS AND HABITAT** Open forests and plantations; sometimes in human habitations around forest edges (<600m asl). Arboreal; can make extended leaps. Diet comprises lizards. Breeding habits are unstudied.

Philippine Rat Snake ■ *Coelognathus erythrurus* 167cm

DESCRIPTION Top of body is brown to olive; a narrow, dark post-ocular stripe runs

to angle of jaws; posterior third of body is darker; tail is reddish brown; belly is plain cream or white; iris is golden brown. Body is slender; head is elongate and distinct from neck; eyes are large with a rounded pupil; tail is long and slender; dorsals are keeled. **DISTRIBUTION** Borneo. Extralimitally: Palawan, Sulu Archipelago, Sulawesi, Mindanao, Samar, Leyte, Negros, Mindoro, Luzon and other Philippines islands. **HABITS AND HABITAT** Lowlands to mid-hills, including forest edges and cultivated areas (<850m asl). Terrestrial. Diet comprises small rodents, birds and lizards. Oviparous, clutches comprising 6–10 eggs (size unknown).

Yellow-striped Rat Snake ■ *Coelognathus flavolineatus* 180cm
(Bahasa Malaysia: Ular Laju Ekor Hitam, Ular Sawa. Bahasa Indonesia: Sawa Angin; Ular Babi. Thai: Ngu Tang-ma-prao Dam)

DESCRIPTION Top of body is brownish grey or brownish olive; a dark post-ocular stripe runs to above back of mouth and another along nape; several short, dark stripes or elongated blotches are present on dorsum and flanks; belly is pale yellow anteriorly, some ventrals with dark grey edges; tail is darker posteriorly; subcaudals are dark grey or black. Body is slender; head is nearly distinct from neck; snout is long; eyes are large with a rounded pupil; tail is long and slender; dorsals are keeled; cloacal scute is entire. **DISTRIBUTION** Myanmar, Thailand, Peninsular Malaysia, Singapore, Sumatra, Pulau Nias, Mentawai and Riau archipelagos, Pulau Weh, Pulau Bangka, Pulau Belitung, Borneo, Java. Extralimitally: Cambodia, Vietnam, Andaman Islands, Sulawesi. **HABITS AND HABITAT** Forested lowlands, parks and gardens (<900m asl). Terrestrial but can climb. Diet comprises rodents, birds, frogs and lizards. Oviparous, laying clutches of 5–12 eggs (51–62 × 23–25.5mm).

Copper-head Rat Snake ▪ *Coelognathus radiatus* 230cm
(Bahasa Malaysia: Ular Rusuk Kerbau. Bahasa Indonesia: Ular Racer Berkepala Tembaga.
Mandarin Chinese: Sangianjing She. Thai: Ngu Tang-ma-prao Lai Keet)

DESCRIPTION Top of body is greyish brown or yellowish brown, with 4 (2 broad,
2 narrow) black stripes along anterior of body; a cream stripe runs along upper 2 stripes;
lower stripes are narrower and typically broken up; forehead is coppery brown, with 3 black
lines radiating from eyes; belly is grey to yellowish grey; iris is golden yellow; tongue is dark
brown to violet. Body is slender; head is slightly distinct from neck; eyes are large with a
rounded pupil; tail is long and slender; dorsals are smooth anteriorly and on flanks, weakly
keeled posteriorly; cloacal scute is entire. **DISTRIBUTION** Myanmar, Thailand, Peninsular

Malaysia, Singapore, Sumatra, Pulau
Bangka, Borneo, Java. Extralimitally:
India, Bangladesh, Bhutan, Nepal, China,
Cambodia, Laos, Vietnam. **HABITS
AND HABITAT** Open grasslands and
light forests from lowland plains to sub-
montane limits (<1,400m asl). Diet
comprises frogs, birds and rats. Oviparous,
clutches comprising 5–23 eggs (40–53 ×
20–26mm).

Mountain Dwarf Snake ▪ *Collorhabdium williamsoni* 30cm
DESCRIPTION Back of body light brown, with five dark stripes; no pattern on neck
region; two yellow square-shaped marks behind parietals; belly yellow. **DISTRIBUTION**
Peninsular Malaysia (Cameron Highlands, Pahang State and Bukit Larut, Perak State).
HABITS AND HABITAT Submontane and montane forests at elevations of 1,036–
1,829m asl. Diet unknown, suspected to be worms and arthropods. Oviparous, clutch size
unknown. Eggs elongate, measuring 23mm.

Stripe-tailed Bronzeback Tree Snake
■ *Dendrelaphis caudolineatus* 152cm
(Bahasa Malaysia/Indonesia: Ular Padang. Iban: Ular Bendira. Thai: Ngu Sai-man Lang Lai)

DESCRIPTION Top of body is olive-brown; a pale green stripe runs along lower sides of body, and is edged dorsally by a narrow black stripe and ventrally by a broad black stripe, these stripes most distinct on tail; belly is pale green. Body is slender; head is distinct from neck; snout is bluntly rounded; tail is long; eyes are large with a rounded pupil; dorsals are smooth; ventrals and subcaudals have sharp keels on outer edges; cloacal scute is divided.
DISTRIBUTION Myanmar; Thailand; Peninsular Malaysia; Singapore; Sumatra; Pulau Babi; Pulau Bangka; Batu, Natuna, Mentawai and Riau archipelagos; Pulau Belitung, Pulau Nias; Borneo. Extralimitally: Laos, Palawan, Luzon, Camiguin and Sulu archipelagos.

HABITS AND HABITAT Lowland and sub-montane forests, and parks and gardens (<1,524m asl). Arboreal. Diet includes frogs and lizards. Oviparous, clutches comprising 5–8 eggs (12 × 48mm).

Blue Bronzeback Tree Snake ■ *Dendrelaphis cyanochloris* 143cm
(Bahasa Malaysia: Ular Tembaga Wall. Thai: Ngu Sai-man Fa Kieo)

DESCRIPTION Top of body is olive, the scales edged with black, and with a broad black temporal stripe extending to beyond level of neck, where it breaks up into spots; during threat display, blue interstitial skin is exposed, the resulting effect being blue with brown-barred neck and body; all dorsals, except 1st row, have black anterior and lower edges; there is no pale ventro-lateral stripe; belly is yellow. Body is slender; head is distinct from neck; eyes are small with a rounded pupil; vertebrals are enlarged; dorsals are smooth; cloacal scute is divided.

DISTRIBUTION Myanmar, Thailand, Peninsular Malaysia. Extralimitally: Bangladesh, Bhutan, India, Laos. **HABITS AND HABITAT** Lowland forests and agricultural areas. Arboreal. Diet comprises lizards, birds and, possibly, frogs. Oviparous, clutches comprising 3–5 eggs (size unknown).

Beautiful Bronzeback Tree Snake ▪ *Dendrelaphis formosus* 147cm

DESCRIPTION Top of body is blue; a black post-ocular stripe runs from rostral, across temporal region and onto neck; 3 dark lateral stripes are present on posterior third of body; dorsals are dark-edged; no light ventro-lateral stripe is present; subcaudals have a black medial point. Body is slender; head is wide behind snout and distinct from neck; eyes are

large with a rounded pupil; vertebrals are enlarged; tail is slender; cloacal scute is divided. **DISTRIBUTION** Thailand, Peninsular Malaysia, Sumatra, Mentawai Archipelago, Pulau Bangka, Pulau Belitung, Borneo, Java. **HABITS AND HABITAT** Lowland forests and open areas. Arboreal. Diet comprises frogs and lizards. Oviparous, clutches comprising 6–8 eggs (31–42.5 × 11.5–13mm).

Haas's Bronzeback Tree Snake ▪ *Dendrelaphis haasi* 94.5cm

DESCRIPTION Top of body is olive-brown; a narrow post-ocular stripe covers ventral part of lower temporals and ends at back of jaws or extends onto neck; oblique black bars are present on sides of neck; a pale ventro-lateral stripe is present but is not bordered by black lines; belly is light yellow or green. Body is slender; head is distinct from neck; eyes

are small with a rounded pupil; tail is long; vertebrals are greatly enlarged, being larger than 1st row of dorsals; dorsals are smooth; cloacal scute is divided. **DISTRIBUTION** Peninsular Malaysia, Sumatra, Pulau Nias, Pulau Belitung, Mentawai Archipelago, Borneo, Java. **HABITS AND HABITAT** Lowland forests. Diet and reproductive habits are unstudied.

Kopstein's Bronzeback Tree Snake ■ *Dendrelaphis kopsteini* 142.5cm
(Bahasa Brunei: Ular Kasau)

DESCRIPTION Top of body is bronze-brown; a black post-ocular stripe runs across lower half of temporal region to end of jaws; vertebrals have a broad black posterior margin; interstitial skin on anterior of body is brick-red; belly is grey. Body is slender; head is distinct from neck; 2 supralabials touch orbit of eye; eyes are moderate in size with a rounded pupil; vertebrals are larger than lowest dorsal row; cloacal scute is divided. **DISTRIBUTION** Thailand, Peninsular Malaysia, Singapore, Sumatra, Mentawai Archipelago, Borneo. **HABITS AND HABITAT** Lowland dipterocarp forests (<700m asl). Arboreal, living in shrubs and on low tree branches. Diet includes geckos. Oviparous, clutches comprising 8 eggs (size unknown).

Painted Bronzeback Tree Snake ■ *Dendrelaphis pictus* 125cm
(Bahasa Malaysia: Ular Lidi. Bahasa Indonesia: Ular Tampar Jawa; Ular Tali. Iban: Ular Meresian. Thai: Ngu Sai-man Pra Inthra)

DESCRIPTION Top of body is bronze-brown or brownish olive; a black-edged yellow or cream ventro-lateral stripe runs along flanks; forehead is brown with a black post-ocular stripe that covers more than half temporal region and extends to neck; a blue or greenish-blue neck patch is displayed when snake is excited. Body is slender; head is distinct from neck; eyes are large with a rounded pupil; vertebrals are smaller than or equal to 1st row of dorsals; dorsals are smooth; cloacal scute is divided. **DISTRIBUTION** Thailand, Peninsular Malaysia, Singapore, Sumatra, Pulau Belitung, Mentawai Archipelago, Borneo, Java, Bali. Extralimitally: China, Laos, Cambodia, Vietnam, Philippines. **HABITS AND HABITAT** Lowland and sub-montane forests; also parks, gardens, plantations and human habitation (<1,524m asl). Arboreal. Diet comprises frogs and lizards. Oviparous, clutches comprising 3–8 eggs (22–38.5 × 8.5–11mm).

Northwestern Painted Bronzeback Tree Snake
■ *Dendrelaphis proarchos* 121cm

DESCRIPTION Top of body is dark brown; a yellow ventro-lateral stripe, edged with 1–2 black lines, runs along flanks; forehead is brown with a black post-ocular stripe that covers

most of temporal region and extends to neck; a blue neck patch is displayed when snake is excited. Body is slender; head is distinct from neck; eyes are large with a rounded pupil; vertebrals are smaller than or equal to 1st row of dorsals; dorsals are smooth; cloacal scute is entire or divided. **DISTRIBUTION** Myanmar. Extralimitally: E India, Bangladesh, Laos, Vietnam. **HABITS AND HABITAT** Lowland forests. Arboreal. Diet includes frogs and lizards. Oviparous, clutches comprising 7–8 eggs (12 × 41mm).

Striated Bronzeback Tree Snake ■ *Dendrelaphis striatus* 102cm
(Thai: Ngu Sai-man Lai Chieng)

DESCRIPTION Top of body is bronze-brown; labials and gular region are yellow; a narrow, dark stripe runs between nostril to orbit; a dark post-ocular stripe covers temporal region

and extends onto neck; neck is yellow when inflated; black oblique bars are present on sides of body. Body is slender; head is distinct from neck; snout is short and rounded; eyes are moderate in size with a rounded pupil; vertebrals are slightly enlarged; dorsals are smooth; cloacal scute is divided. **DISTRIBUTION** Thailand, Peninsular Malaysia, Sumatra, Borneo. **HABITS AND HABITAT** Lowland forests. Arboreal. Diet and reproductive habits are unstudied.

Mountain Bronzeback Tree Snake ■ *Dendrelaphis subocularis* 88cm
(Thai: Ngu Sai-man Kled Tai Ta Yai)

DESCRIPTION Top of body is bronze-brown, the scales with black edges; lower flanks beyond scale row 2 are bright cream or greenish white; head and neck are olive-green; a

dark post-ocular stripe extends to sides of neck, breaking up thereafter into bars; gulars are sometimes yellow. Body is slender; head is distinct from neck; snout is rounded; eyes are small with a rounded pupil; dorsals are smooth; vertebrals are weakly enlarged; cloacal scute is divided. **DISTRIBUTION** Myanmar, Thailand. Extralimitally: S China, Laos, Cambodia, Vietnam. **HABITS AND HABITAT** Mid-hill forests (altitude unknown). Arboreal. Diet and reproductive habits are unstudied.

Indian Bronzeback Tree Snake ■ *Dendrelaphis tristis* 150cm

DESCRIPTION Top of body is plain bronzy brown or sometimes purplish brown; vertebrals on neck and forebody are yellow; a buff flank stripe runs from neck to vent; a pale blue patch on neck between scales is displayed when snake is excited; belly is pale grey, green or yellow; iris is golden. Body is slender; head is distinct from neck; eyes are large with a

rounded pupil; tail is about a third snout–vent length; dorsals are smooth, with apical pits; cloacal scute is divided. **DISTRIBUTION** Myanmar. Extralimitally: India, Pakistan, Bangladesh, Sri Lanka, Nepal. **HABITS AND HABITAT** Forest edges and around human habitation (<2,000m asl). Arboreal but may forage on land. Diet comprises frogs, lizards, bird eggs and insects. Oviparous, laying 6 eggs (29–39 × 10–12mm).

Davison's Bridled Snake ■ *Dryocalamus davisonii* 92cm

DESCRIPTION Back of body black, with pale green or white cross-bars, numbering 29–31 on body and 19–21 on tail; bars widest anteriorly, and widen on flanks; at back,

they form a reticulate design; back of forehead with a dark median stripe; lips white; belly unpatterned white; dorsals are smooth; cloacal scales entire. **DISTRIBUTION** Southern Myanmar (Tenasserim, Tanintharyi Division) and Thailand. Extralimitally: Laos, Cambodia and Vietnam. **HABITS AND HABITAT** Lowland temperate and subtropical forests, including evergreen and moist deciduous forests, and grasslands, up to submontane limits at 1,000m asl. Diet includes lizards. Clutches comprise 3–4 eggs, measuring 35 x 9mm.

Half-banded Bridled Snake ■ *Dryocalamus subannulatus* 60cm
(Thai: Ngu Plong-chanuan Malayu)

DESCRIPTION Top of body is tan or light brown, with large brown transverse spots; on flanks, smaller spots are present; 2 dark post-ocular streaks are present; belly is plain yellow. Body is slender and compressed; head is depressed and distinct from neck; eyes are large with a vertical pupil; dorsals are smooth; cloacal scute is entire. **DISTRIBUTION** Myanmar, Thailand, Peninsular Malaysia, Singapore, Sumatra, Mentawai and Riau archipelagos,

Borneo. Extralimitally: Palawan. **HABITS AND HABITAT** Lowland forests and disturbed areas (altitude unknown). Terrestrial and arboreal. Diet comprises small vertebrates. Reproductive biology is unstudied.

Three-banded Bridled Snake ■ *Dryocalamus tristrigatus* 65cm

DESCRIPTION Top of body is dark brown with 3 white stripes; forehead shields are white-edged; supralabials are white; belly is cream. Body is slender and compressed; head is depressed and distinct from neck; eyes are large with a vertical pupil; dorsals are smooth; cloacal scute is entire. **DISTRIBUTION** Borneo, Natuna Archipelago. Extralimitally: Balabac, Palawan. **HABITS AND HABITAT** Lowland forests, on rocky biotopes and trees (altitude unknown). Diet includes lizards. Breeding habits are unstudied.

Keel-bellied Whip Snake ■ *Dryophiops rubescens* 75cm
(Thai: Ngu Tao)

DESCRIPTION Top of body is reddish brown with small black spots; forehead has dark streaks; a dark post-ocular streak is present; labials have dark spots; belly is yellow or olive.

Body is slender and compressed; head is distinct from neck; eyes are large with a horizontal pupil; tail is long and slender; dorsals are smooth; cloacal scute is divided. **DISTRIBUTION** Thailand, Peninsular Malaysia, Singapore, Sumatra, Mentawai and Natuna archipelagos, Borneo, Java. Extralimitally: Cambodia. **HABITS AND HABITAT** Lowland forests and forest edges (<300m asl). Arboreal, on shrubs and low branches of trees. Diet comprises lizards. Oviparous, clutches comprising 2–3 eggs (numbers unknown).

Eastern Rat Snake ■ *Elaphe cantoris* 200cm

DESCRIPTION Top of body is greyish brown to yellowish brown; dark or light borders to scales produce a reticulate pattern; dark brown or reddish-brown blotches on dorsum form transverse bands towards posterior of body; smaller blotches are present on flanks; gular region is yellow or orange; belly is yellow, turning pink towards tail; iris is red. Body is slender but robust; head is elongate and slightly distinct from neck; eyes are large with

a rounded pupil; ventral keel is distinct; dorsals are smooth or weakly keeled; cloacal scute is entire. **DISTRIBUTION** Myanmar. Extralimitally: E India, Bhutan, Nepal. **HABITS AND HABITAT** Montane and sub-montane forests (1,000–2,300m asl). Diet is likely to include small mammals and birds. Oviparous, producing clutches of 5–10 eggs (size unknown).

Mandarin Rat Snake ■ *Euprepiophis mandarinus* 170cm
(Mandarin Chinese: Gaosha She)

DESCRIPTION Top of body is grey to greyish brown, dorsals with brownish-red centres; dorsum and tail have large, rounded yellow blotches, edged with black and yellow; forehead has a dark V-shaped pattern; belly is cream, sometimes with large black blotches. Body is robust; head is short and slightly distinct from neck; snout is obtuse; tail is short and stout; eyes are small with a rounded pupil; dorsals are smooth; cloacal scute is divided. **DISTRIBUTION** Myanmar. Extralimitally: Vietnam, E India, China, possibly Laos. **HABITS AND HABITAT** Plains to montane regions, in open forests with rocky substrate, scrubland and agricultural fields (500–3,000m asl). Terrestrial, near water. Diet comprises mice and shrews. Oviparous, clutch comprising 2–10 eggs (50–57 × 23–27mm).

Orange-bellied Snake ■ *Gongylosoma baliodeirum* 45cm
(Bahasa Indonesia: Ular Tanah Bertotol-totol. Kelabit: Depong. Thai: Ngu Sai-tong Lai Taeb)

DESCRIPTION Top of body is dark brown to reddish brown, with paired rows of cream spots; upper labials are edged with dark grey; belly is yellowish cream, sometimes with fine dark spots. Body is slender and cylindrical; head is slightly wider than neck; nasals are divided; eyes are small with a rounded pupil; dorsals are smooth; cloacal scute is divided. **DISTRIBUTION** Thailand, Peninsular Malaysia, Singapore, Sumatra, Pulau Nias, Natuna Archipelago, Borneo, Java. **HABITS AND HABITAT** Lowland to sub-montane forests (<1,525m asl). Terrestrial. Diet comprises spiders, other arthropods and lizards. Oviparous, clutches comprising 2–3 eggs (22–24 × 7.5mm).

Pulau Tioman Ground Snake ■ *Gongylosoma mukutense* 42.9cm

DESCRIPTION Top of body is brick-red anteriorly, fading to brown-grey posteriorly; forehead is brown; supralabials are white, edged with black; nuchal band is confluent with vertebral stripe; a white post-ocular patch is present; remnants of 5 thin white stripes are

visible anteriorly; belly is cream. Body is slender; head is wider than neck; snout is short and rounded; eyes are large with a rounded pupil; tail is long; dorsals are smooth; cloacal scute is divided. **DISTRIBUTION** Pulau Tioman, Peninsular Malaysia. **HABITS AND HABITAT** Coastal forests (10m asl). Terrestrial. Diet and reproductive habits are unstudied.

Khasi Hills Trinket Snake ■ *Gonyosoma frenatum* 150cm

DESCRIPTION Top of body is grass-green to olive; supralabials and infralabials are light green; a black post-ocular stripe runs to angle of jaws; belly is pale green to white, darkening posteriorly; iris is golden yellow. Body is slender and compressed; snout is elongate, its tip pointed and slightly arched forward; eyes are large with a rounded pupil; tail is long and prehensile, tapering to a point; ventral keels are developed; dorsals

are weakly keeled, except outer 2–3 rows; cloacal scute is divided. **DISTRIBUTION** Myanmar. Extralimitally: Vietnam, E India, E China. **HABITS AND HABITAT** Evergreen and sub-montane forests (550–2,000m asl). Arboreal, in shrubs, low trees and bamboo groves. Diet comprises lizards, rats and birds. Oviparous, clutches comprising 9 eggs (28 × 15mm).

Royal Tree Snake ■ *Gonyosoma margaritatus* 200cm

DESCRIPTION Top of body is bright green, each scale edged with black; 5 orangish-red bands are present on posterior of body and tail; a black post-ocular stripe is present; forehead has black streaks; belly is yellowish pink. Body is robust, elongate and compressed; head is distinct from neck; snout is elongate and squarish; eyes are large with a rounded pupil; tail is long and tapering; dorsals are smooth; cloacal scute is divided. **DISTRIBUTION** Peninsular Malaysia, Singapore, Borneo. **HABITS AND HABITAT** Lowland forests (<700m asl). Arboreal. Diet and reproductive habits are unstudied.

Red-tailed Racer ■ *Gonyosoma oxycephalum* 240cm
(Bahasa Malaysia: Ular Laju Ekor Merah, Ular Pucuk, Ular Selenseng. Bahasa Indonesia: Ular Bangka Laut, Ular Hidjau, Ular Racer Berekor Merah. Iban: Ular Mati Iko, Ular Mati Elok. Thai: Ngu Kieo Kab-mak)

DESCRIPTION In adults, top of body is emerald-green or light green, with a light green throat and a black stripe from nostril to above level of upper jaw; tail tip is yellowish brown or reddish orange; belly is yellow; tongue is bluish black. Juveniles are olive-brown with narrow white bars towards posterior. Body is slender, elongate and compressed; head is distinct from neck; snout is elongate and squarish; eyes are large with a rounded pupil; tail is long and tapering; dorsals are smooth or weakly keeled; cloacal scute is divided. **DISTRIBUTION** Myanmar; Thailand; Peninsular Malaysia; Singapore; Sumatra; Pulau Nias; Pulau Bangka; Pulau Belitung; Riau, Mentawai and Natuna archipelagos; Borneo; Java. Extralimitally: Laos, Cambodia, Vietnam, Andaman Islands, Lombok, Sulawesi, Balabac, Bohol, Bongao, Dinagat, Lubang, Luzon, Mindoro, Negros, Palawan. **HABITS AND HABITAT** Lowland tropical and subtropical forests, including mangrove swamps, plantations and gardens (<750m asl). Arboreal and terrestrial. Diet comprises rodents and birds. Oviparous, clutches comprising 5–12 eggs (65mm diameter).

Green Trinket Snake ▪ *Gonyosoma prasinum* 120cm
(Mandarin Chinese: Lujing She. Thai: Ngu Tang-ma-prao Kieo)

DESCRIPTION Top of body is uniformly green or turquoise; tail tip is brown; labials are green or yellowish green; belly is pale green; has a faint dark post-ocular stripe; iris

is yellow; tongue is reddish brown. Body is slender; snout is long; head is slightly distinct from neck; eyes are large with a rounded pupil; dorsals are weakly keeled, except outer 2–3 rows; ventral keels are well developed; tail is long and slender; cloacal scute can be entire or divided. **DISTRIBUTION** Myanmar, Thailand, Peninsular Malaysia. Extralimitally: Laos, Vietnam, E India, China. **HABITS AND HABITAT** Submontane and montane forests (80–2,650m asl). Arboreal, on bamboo; also roofs of thatched houses. Diet includes lizards, mammals and birds. Oviparous, producing clutches of 5–9 eggs (31–39 × 16–19mm).

Stripe-necked Snake ▪ *Liopeltis frenatus* 76cm

DESCRIPTION Top of body is olive, scales edged with black (sometimes also with white), forming longitudinal stripes on anterior half of body; a broad black post-ocular stripe extends to neck; supralabials and belly are cream; tongue is orange. Body is slender and

cylindrical; head is distinct from neck; snout does not project; eyes are large with a rounded pupil; dorsals are smooth; tail is long; cloacal scute is divided. **DISTRIBUTION** Myanmar and Thailand. Extralimitally: Laos, Vietnam, E India. **HABITS AND HABITAT** Subtropical and montane forests (600–1,830m asl). Terrestrial. Diet probably includes frogs. Oviparous, laying clutches of 4–5 eggs (25–28 × 7–9mm).

Stoliczka's Ringneck ■ *Liopeltis stoliczkae* 60cm

DESCRIPTION Top of body is brown or greyish brown, with a broad black stripe on sides of head, extending to anterior body and fading thereafter; has a grey stripe on outer margins of ventrals; belly is pale grey. Body is slender and cylindrical; head is distinct from neck and depressed; snout is projecting, being twice as long as diameter of orbit; eyes are large with a rounded pupil; tail is long and slender; dorsals are smooth; cloacal scute is divided. **DISTRIBUTION** Myanmar. Extralimitally: Laos, Cambodia, E India. **HABITS AND HABITAT** Evergreen and deciduous forested mid-hills (<700m asl). Arboreal, on bamboo. Diet and reproductive habits are unstudied.

Tricoloured Ringneck ■ *Liopeltis tricolor* 56cm
(Thai: Ngu Sai-tong Malayu)

DESCRIPTION Top of body is yellowish olive; a dark post-ocular streak extends beyond neck; belly is yellowish cream, with an olive streak on sides of each scale. Body is slender; head is indistinct from neck; snout is long; eyes are large with a rounded pupil; tail is long; dorsals are smooth; cloacal scute is divided. **DISTRIBUTION** Thailand, Peninsular Malaysia, Singapore, Sumatra, Mentawai Archipelago, Borneo, Java. Extralimitally: Vietnam, Palawan. **HABITS AND HABITAT** Lowland forests (altitude unknown). Arboreal, on small trees and other low vegetation. Diet comprises insects and spiders. Reproductive biology is unstudied.

Dusky Wolf Snake ■ *Lycodon albofuscus* 207cm
(Thai: Ngu Plong-chanuan Borneo)

DESCRIPTION In adults, top of body is plain dark brown or brownish black; supralabials are yellow; belly is plain cream or yellow. Juveniles have 30–40 narrow white or yellow bands on dorsum. Body is slender and sub-cylindrical; head is wider than neck; snout is short, blunt and depressed; eyes are small with a vertical pupil; tail is long; dorsals are

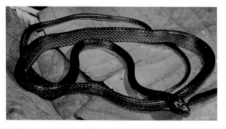

keeled; cloacal scute is divided. **DISTRIBUTION** Thailand, Peninsular Malaysia, Sumatra, Pulau Nias and Borneo. **HABITS AND HABITAT** Open lowland forests and forest edges, usually with streams (<500m asl). Terrestrial, with some arboreal activities. Diet comprises lizards and frogs. Reproductive habits are unstudied.

Indian Wolf Snake ■ *Lycodon aulicus* 80cm

DESCRIPTION Top of body is brown or greyish-brown, with 12–19 white cross-bars, that are sometimes speckled with brown, expanding laterally to enclose triangular patches; belly is cream coloured or yellowish-white. Body is slender and sub-cylindrical; head is flattened; snout projects beyond lower jaw; eyes are small with a vertical pupil; tail is long; dorsals are smooth; cloacal scute is divided. **DISTRIBUTION** Myanmar. Extralimitally: India, Bangladesh, Nepal, Sri Lanka, possibly Maldive Archipelago. **HABITS AND HABITAT** Lowlands, sometimes inside human habitation (<500m asl). Arboreal. Diet comprises geckos, snakes and rodents. Oviparous, producing 3–11 eggs (diameter 25–32mm).

Butler's Wolf Snake ▪ *Lycodon butleri* 100cm
(Thai: Ngu Plong-chanuan Malayu)

DESCRIPTION In adults, top of body is dark bluish grey or blackish brown, suffused with dark pigments; belly is cream with dark brown cross-bars, these most distinct at mid-body. In juveniles, top of body has 40–50 irregular cross-bars. Body is slender and sub-cylindrical; head is flattened; eyes are small with a vertical pupil; tail is long; dorsals are weakly keeled; cloacal scute is entire. **DISTRIBUTION** Peninsular Malaysia. **HABITS AND HABITAT** Montane forests and adjacent human habitations (1,220–2,031m asl). Arboreal. Diet comprises geckos. Reproductive habits are unstudied.

Island Wolf Snake ▪ *Lycodon capucinus* 76cm
(Bahasa Malaysia: Ular Rumah Biasa. Bahasa Indonesia: Ular Cecak, Ular Tanah. Thai: Ngu Soi-luang)

DESCRIPTION Top of body is brown, grey-brown or purple, with a narrow yellow or cream band at back of head that may be brown-spotted; interstitial skin is yellow or grey; scales of body are light-edged and form indistinct cross-bars or a reticulate pattern; belly is cream or light yellow. Body is slender and sub-cylindrical; snout is rounded; head is flattened; eyes are small with a vertical pupil; tail is long; dorsals are smooth; cloacal scute is entire. **DISTRIBUTION** Myanmar, Thailand, Peninsular Malaysia, Singapore, Sumatra, Borneo, Java, Bali. Extralimitally: Andaman Islands, Vietnam, Sulawesi, Lesser Sundas, China, Philippines. **HABITS AND HABITAT** Lowland forests and mid-hills (<600m asl), including human habitation. Terrestrial and arboreal. Diet includes lizards. Oviparous, producing 3–11 eggs (20–30 × 10mm).

Cardamom Mountains Wolf Snake ■ *Lycodon cardamomensis* 31.6cm

DESCRIPTION Top of body is black with 12 white bands on body and 6 on tail, these wider on flanks (5–9 scales wide) than on vertebral region (3–5 scales wide); forehead

is black, except for pale suture between frontals and parietals; belly is mostly cream. Body is slender and sub-cylindrical; head is flattened; eyes are small with a vertical pupil; tail is long; dorsals are weakly keeled; cloacal scute is entire. **DISTRIBUTION** Thailand. Extralimitally: Cambodia. **HABITS AND HABITAT** Lowland forests (500m asl). Terrestrial. Diet and reproductive biology are unstudied.

Brown Wolf Snake ■ *Lycodon effraenis* 100cm
(Thai: Ngu Plong-chanuan Malayu)

DESCRIPTION In adults, top of body is reddish brown or dark brown; belly is unpatterned brown. Juveniles have 3 broad yellow rings encircling body, and yellow streaks on sides of head or a distinct canthal stripe. Body is slender and sub-cylindrical; head is flattened; snout is rounded; eyes are small with a vertical pupil; tail is long; dorsals are smooth or weakly

keeled; cloacal scute is entire. **DISTRIBUTION** Peninsular Malaysia, Sumatra, Borneo. **HABITS AND HABITAT** Lowland forests and disturbed habitats (<700m asl). Terrestrial and arboreal. Diet is unstudied but presumably comprises lizards and snakes. Reproductive habits are unstudied.

Banded Wolf Snake ■ *Lycodon fasciatus* 85cm
(Thai: Ngu Plong-chanuan Moeng Noe)

DESCRIPTION Top of body is glossy black, with 22–48 irregular cross-bars on body and tail, or a reticulate or spotted pattern; belly is blotched; iris is flecked with grey. Body is

slender and sub-cylindrical; head is flattened; eyes are small with a vertical pupil; tail is long; dorsals are weakly keeled, the keels more pronounced posteriorly; cloacal scute is entire. **DISTRIBUTION** Myanmar, Thailand. Extralimitally: Laos, Vietnam, India, Bhutan, China. **HABITS AND HABITAT** Temperate evergreen forests from mid-hills to montane limits (914–2,300m asl). Arboreal, on trees and bushes. Diet comprises snakes and lizards. Oviparous, producing clutches of 4–14 eggs (size unknown).

Yellow-speckled Wolf Snake ■ *Lycodon jara* 55cm

DESCRIPTION In adults, top of body is brown or purplish black, finely stippled throughout with paired yellowish-white spots or short longitudinal lines on each scale; supralabials and belly are plain white. Juveniles show a white or yellow collar. Body is slender and sub-cylindrical; head is flattened, and weakly or not projecting beyond lower jaw; eyes are small with a vertical pupil; tail is long; dorsals are smooth; ventrals are

not angular laterally; subcaudals are paired; cloacal scute is divided. **DISTRIBUTION** Myanmar. Extralimitally: Nepal, Bangladesh, E India. **HABITS AND HABITAT** Open forests with bushes, and agricultural areas. Diet comprises frogs, lizards and rodents. Oviparous (egg numbers and size unknown).

Laos Wolf Snake ■ *Lycodon laoensis* 50cm
(Thai: Ngu Plong-chanuan Lao)

DESCRIPTION Top of body is shiny black with white-edged yellow cross-bars, these numbering 13–29 on body and 8–18 on tail, and becoming narrower towards posterior; forehead and labials are deep blue; belly is plain cream. Body is slender and sub-cylindrical;

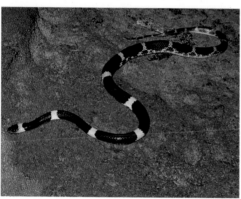

head is flattened; eyes are small with a vertical pupil; tail is long; dorsals are smooth; cloacal scute is divided. **DISTRIBUTION** Thailand, Peninsular Malaysia. Extralimitally: Laos, Cambodia, Vietnam, E India, China. **HABITS AND HABITAT** Evergreen forests in plains and low hills (altitude unknown). Arboreal, climbing trees up to 9m. Diet includes frogs and lizards. Oviparous, clutches containing 5 eggs (size unknown).

Snake-eating Wolf Snake ■ *Lycodon ophiophagus* 90.9cm

DESCRIPTION Top of body is brownish black to nearly black, with 20–21 white bands on body and 14 on tail; belly is whitish cream anteriorly, marked with numerous dark brown blotches posteriorly. Body is slender and elongate; head is flattened and distinct from

neck; eyes are moderate in size with a vertical pupil; tail is long; dorsals are smooth; ventrals are keeled; cloacal scute is divided. **DISTRIBUTION** Peninsular Thailand. **HABITS AND HABITAT** Lowland forests of the provinces of Chumphon, Ranong and Phang Nga, along forest streams. Diet in the wild is unknown, but in captivity accepts snakes, frogs and fish. Reproductive habits are unstudied.

Northern Wolf Snake ■ *Lycodon septentrionalis* 118cm

DESCRIPTION Back of body purplish-black, with 20–35 white transverse bands and 10–17 on tail, that widen on flanks; belly white, sometimes spotted or barred with

black; undersurface of tail with black speckling. **DISTRIBUTION** Myanmar (Kayin, Kachin and northern Mon states) and northern Thailand (Chiang Mai Province). Extralimitally: eastern India and southern China, northern Laos, Cambodia and Vietnam. **HABITS AND HABITAT** Mid-hills of evergreen forests, between 220–500m asl. Diet comprises frogs and lizards. Reproductive habits unstudied.

White-banded Wolf Snake ■ *Lycodon subcinctus* 102cm
(Bahasa Indonesia: Ular Cecak Belang. Thai: Ngu Plong-chanuan Ban)

DESCRIPTION Top of body is black or dark brown, bearing 9–15 cream bands that are 3–5 scales wide; pattern is more distinctive in juveniles, fading with growth; in adults, remains of bands persist into the grey or cream belly. Body is slender; head is flattened; snout is rounded; eyes are small with a vertical pupil; tail is long; dorsals bear weak keels; cloacal scute is divided. **DISTRIBUTION** Myanmar, Thailand, Peninsular Malaysia, Singapore, Sumatra,

Pulau Nias, Mentawai Archipelago, Borneo, Java. Extralimitally: Nicobar Archipelago, Laos, Cambodia, Vietnam, Lombok, Sumbawa, Philippines, China. **HABITS AND HABITAT** Lowland forests and mid-hills (<1,000m asl). Arboreal and terrestrial. Diet includes geckos and skinks. Oviparous, laying 5–11 eggs (32–36 × 12.5–13mm).

Zaw's Wolf Snake ■ *Lycodon zawi* 48cm

DESCRIPTION Top of body is brownish black, with narrow cream bands that are less distinct posteriorly; labials are pale brown; belly is cream, edges of ventrals with a dark edge; iris is black. Body is slender; head is flattened and distinct from neck; snout is

projecting; eyes are small with a vertical pupil; tail is long; dorsals are smooth; cloacal scute is divided. **DISTRIBUTION** Myanmar. Extralimitally: India. **HABITS AND HABITAT** Both dry and moist deciduous, semi-evergreen and tropical evergreen forests, near streams (<500m asl). Terrestrial. Diet comprises skinks. Reproductive habits are unstudied.

Chan-ard's Mountain Reed Snake ■ *Macrocalamus chanardi* 26.3cm

DESCRIPTION Top of body is chestnut-brown or pale brown, anteriorly with a row of yellow or ochre stripes composed of dark-edged ocelli, or sometimes small black dots; has a dark ventro-lateral stripe edged with a yellow or cream stripe; has 2–4 yellowish-ochre

oblique bars between neck and ventrals; belly is red, pink or orange. Body is robust and sub-cylindrical; head is small; eyes are small with a rounded pupil; tail is short and tapering; dorsals are smooth; cloacal scute is entire. **DISTRIBUTION** Peninsular Malaysia. **HABITS AND HABITAT** Sub-montane forests (1,110–1,500m asl). Sub-fossorial, under fallen logs. Diet comprises earthworms, slugs, and insects and their larvae. Reproductive habits are unstudied.

Golden-bellied Reed Snake ■ *Macrocalamus emas* 21.8cm

DESCRIPTION Back of body dark greyish brown, with row of discontinuous, pale brown eye-like spots on each side; forehead similarly coloured, snout slightly paler; beige stripe along sides of head; belly bright yellow, bordered by broad, dark brown stripes on outer edges of scales; dorsals are smooth; caudal scale is entire. **DISTRIBUTION** Endemic to Peninsular Malaysia (Gunung Brinchang, Cameron Highlands). **HABITS AND HABITAT** Upper montane forests, at elevations above 1,800m. Diet includes earthworms. Reproductive habits unstudied.

Genting Highlands Reed Snake ■ *Macrocalamus gentingensis* 38cm

DESCRIPTION Back of body iridescent black with scattered yellow patches on sides of nape; belly black with a narrow median yellow stripe; dorsals are smooth; caudal scale is entire. **DISTRIBUTION** Endemic to Peninsular Malaysia (Genting Highlands, Pahang State). **HABITS AND HABITAT** Inhabits lower montane, oak-laurel forests, between 1,689–1,181m asl. Diet and reproductive habits unstudied.

Striped Reed Snake ■ *Macrocalamus lateralis* 29.7cm

DESCRIPTION Top of body pale brown or yellowish-brown; sides of anterior body with pale brown eye-like pattern; paired, dark brown ventro-lateral stripes, separated by pale line; belly red, pink or orange, scales on belly edged with dark brown; tail sometimes with a dark median subcaudal line; dorsals are smooth; caudal scale is entire. **DISTRIBUTION**

Southern Thailand and Peninsular Malaysia (Genting Highlands, Bukit Fraser, Cameron Highlands and Kuala Tahan, Bukit Larut and Pinang State). **HABITS AND HABITAT** Hill dipterocarp forests, 400m asl, up to submontane forests, at 1,920m asl. Diet includes earthworms and slugs. Reproductive habits unstudied.

Schulz's Reed Snake ■ *Macrocalamus schulzi* 39.9cm

DESCRIPTION Top of body is mid-brown; lacks ventro-lateral stripes; some dorsal scales are paler anteriorly, darker posteriorly; outer dorsal scale rows are pale yellow, with brown

mottling below; forehead is brown with a pale temporal streak; belly is bright yellow. Body is robust and sub-cylindrical; head is small; eyes are small with a rounded pupil; tail is short and tapering; dorsals are smooth; cloacal scute is entire. **DISTRIBUTION** Peninsular Malaysia. **HABITS AND HABITAT** Sub-montane and montane forests (1,000–1,800m asl). Sub-fossorial. Diet and reproductive habits are unstudied.

Tweedie's Reed Snake ■ *Macrocalamus tweediei* 50cm

DESCRIPTION Top of body is uniformly black; head has a lateral yellow marking extending ventrally; supralabials and infralabials are yellow; belly is chequered black and yellow; tail may have a median subcaudal line. Body is robust and sub-cylindrical; head is small; eyes are small with a rounded pupil; tail is short and tapering; dorsals are smooth; cloacal scute is entire. **DISTRIBUTION** Peninsular Malaysia. **HABITS AND HABITAT** Sub-montane forests (1,500–1,829m asl). Sub-fossorial, under fallen logs. Diet in the wild is unstudied; in captivity, eats geckos. Reproductive habits are unstudied.

White-barred Kukri Snake ■ *Oligodon albocinctus* 101.5cm

DESCRIPTION Top of body is brownish red, sometimes with black-edged white, yellow or fawn-coloured cross-bars, these numbering 19–27 on body and 4–8 on tail; forehead has a dark stripe from supralabials to orbit, and a dark V-shaped mark; belly is cream, yellow or coral-red, with black areas. Body is robust and sub-cylindrical; head is short; snout is blunt and rounded; eyes are moderate in size with a rounded pupil; dorsals are smooth, lacking apical pits; cloacal scute is entire. **DISTRIBUTION** Myanmar. Extralimitally: Vietnam, E India, Bangladesh, Bhutan, Nepal. **HABITS AND HABITAT** Forests and tea gardens, from mid-hills to montane forests (<c. 2,000m asl). Terrestrial. Diet comprises rodents, frogs, lizards and their eggs, and insects. Oviparous, producing clutches of 3 eggs (size unknown).

Spotted Kukri Snake ■ *Oligodon annulifer* 450mm

DESCRIPTION Top of body dark brown, lighter on flanks, with 20–26 orangish-brown blotches edged with greyish-brown; chevron at back of head; belly cream coloured with

black spots on each side; undersurface of tail brick-red. Body robust and subcylindrical; head short and slightly distinct from neck; snout rounded; eyes moderate; dorsals smooth; cloacal scute divided. **DISTRIBUTION** Borneo. **HABITAT AND HABITS** Found in hill dipterocarp forests at about 120m above sea level. Nocturnal and arboreal, and also active on the ground. Feeds on lizards. Reproductive biology unstudied.

Barron's Kukri Snake ■ *Oligodon barroni* 40.1cm
(Thai: Ngu Pi-kaeo Hua Lai Hua-ji)

DESCRIPTION Top of body is light brown; has a series of 10–14 transverse, black-edged brown blotches on body and 2–3 on tail, with 3 indistinct cross-bars between each spot;

forehead is dark brown, with a broad crescent across eye to labials; belly is coral-red, with squarish black spots. Body is moderately built and sub-cylindrical; head is short and indistinct from neck; eyes are moderate in size with a rounded pupil; dorsals are smooth; cloacal scute is entire. **DISTRIBUTION** Thailand. Extralimitally: Laos, Cambodia, Vietnam. **HABITS AND HABITAT** Lowlands, including low-lying offshore islands (c. 300m asl). Terrestrial. Diet and reproductive habits are unstudied.

Javanese Kukri Snake ■ *Oligodon bitorquatus* 37cm

DESCRIPTION Top of body is dark brown or purple, with small red or yellow spots; median series of larger spots may be present; forehead has dark stripes and yellow bands, including 1 on occipital region; belly is red with black spots or blotches. Body is robust and sub-cylindrical; head is short and indistinct from neck; eyes are small with a rounded pupil; dorsals are smooth; subcaudals are paired; cloacal scute is entire. **DISTRIBUTION** Java, Sumbawa. **HABITS AND HABITAT** Sub-montane forests (1,200–1,524m asl). Terrestrial. Diet is unstudied. Oviparous, clutch comprising 3 eggs (size unknown).

Boo Liat's Kukri Snake ■ *Oligodon booliati* 51cm

DESCRIPTION Top of body and flanks are deep maroon-red; 19–22 indistinct transverse brown bars extend from nape along body, fading towards tail; post-ocular stripe is absent; narrow, dark brown stripes are present on supralabials; belly is salmon-pink and lacks spots. Body is robust and sub-cylindrical; head is short and indistinct from neck; eyes are small with a rounded pupil; dorsals are smooth; cloacal scute is entire. **DISTRIBUTION** Pulau Tioman, Peninsular Malaysia. **HABITS AND HABITAT** Lowland forests. Terrestrial. Diet and reproductive habits are unstudied.

Grey Kukri Snake ■ *Oligodon cinereus* 73cm
(Thai: Ngu Pi-kaeo Lai Kra)

DESCRIPTION Top of body is variable – may be reddish brown or red, plain or with black-edged white or grey cross-bars or reticulations; forehead is plain brown; belly is cream. Body is robust and sub-cylindrical; head is short and indistinct from neck; eyes are moderate in size with a rounded pupil; tail is short and blunt; dorsals are smooth; cloacal scute is entire. **DISTRIBUTION** Myanmar, Thailand, Borneo. Extralimitally: Bhutan, E India, Laos, Cambodia, Vietnam, E China. **HABITS AND HABITAT** Lowland and mid-hill forests (457–700m asl). Diet includes spiders and insects. Oviparous, producing clutches of 4–5 eggs (size unknown).

Pegu Kukri Snake ■ *Oligodon cruentatus* 41cm

DESCRIPTION Top of body is olive-brown, with 4 dark longitudinal stripes, the median rows separated by 3 scale rows and extending to tail tip; a lateral stripe, 2.5 scale rows above ventrals, extends to tail tip; a dark transverse bar before orbit extends as a sub-ocular stripe; has an incomplete dark collar; belly is bright yellow, with large, squarish

black spots; subcaudals are red. Body is robust and sub-cylindrical; head is short and indistinct from neck; eyes are moderate in size with a rounded pupil; dorsals are smooth; subcaudals are paired; cloacal scute is divided. **DISTRIBUTION** Myanmar. **HABITS AND HABITAT** Forested lowlands. Terrestrial. Diet and reproductive biology are unstudied.

Spot-tailed Kukri Snake ■ *Oligodon dorsalis* 62.8cm

DESCRIPTION Top of body is dark brown or purple, with a light vertebral stripe, this sometimes dark-edged or containing small black spots; a 2nd stripe runs along 2nd and 3rd dorsal scale rows; tail dorsum shows large black spots; forehead is dark brown with 2 cross-bars; belly is orange; subcaudals are crimson. Body is robust and sub-cylindrical; head is short and indistinct from neck; eyes are moderate in size with a rounded pupil; dorsals are smooth; cloacal scute is divided. **DISTRIBUTION** Myanmar, Thailand. Extralimitally: E India, Bangladesh, Bhutan. **HABITS AND HABITAT** Evergreen forests from lowlands to montane regions (<1,980m asl). Diet is unstudied. Oviparous, producing at least 2 eggs (size unknown).

Jewelled Kukri Snake ■ *Oligodon everetti* 42cm

DESCRIPTION Top of body is greyish brown with 3 blackish-brown stripes, the broadest (3 scales wide) along vertebral region and enclosing short white and red or orange bars; has a dark stripe on each lower flank that encloses white spots; has a dark, backwards-pointing V-shaped mark on forehead; a 2nd mark present over snout crosses orbit; belly is plain coral-red. Body is of moderate build and sub-cylindrical; head is short and indistinct from neck; eyes are moderate in size with a rounded pupil; dorsals are smooth; cloacal scute is entire. **DISTRIBUTION** Borneo. **HABITS AND HABITAT** Hill dipterocarp forests (<c. 1,000m asl). In leaf litter. Diet comprises skinks, and possibly other vertebrates. Reproductive biology is unknown.

Small-banded Kukri Snake ■ *Oligodon fasciolatus* 88.2cm
(Thai: Ngu Pi-kaeo Lai Taem)

DESCRIPTION Top of body is yellowish olive with 13–18 transverse blotches, these separated by 3–4 dark, wide cross-bars, or with a reticulate pattern; a dark post-ocular stripe does not meet at back of head; has a dorso-lateral stripe along body; tail has a median longitudinal cream band; belly is plain white. Body is robust and sub-cylindrical; head is

short and indistinct from neck; eyes are moderate in size with a rounded pupil; dorsals are smooth; cloacal scute is entire. **DISTRIBUTION** Myanmar, Thailand. Extralimitally: Laos, Cambodia, Vietnam. **HABITS AND HABITAT** Lowland evergreen forests and open areas such as cultivated fields (<c. 260m asl). In captivity, diet comprises frogs and small mammals; diet in the wild unknown. Reproductive habits are unstudied.

Hua Hin Kukri Snake ■ *Oligodon huahin* 55.4cm

DESCRIPTION Top of body pale greyish-brown, with a poorly defined brownish stripe on vertebral and adjacent rows; some vertebral or paravertebral scales, with partly black

posterior margins; forehead surface greyish-brown, with faint, irregular lighter and darker markings; throat and belly ivory-coloured; iris greenish-golden, speckled with black. **DISTRIBUTION** Peninsular Thailand (near Pala-U waterfall, Prachuap Khiri Khan Province). **HABITS AND HABITAT** Bamboo forest. Diet and reproductive habits unstudied.

Unicoloured Kukri Snake ▪ *Oligodon inornatus* 58cm

DESCRIPTION Back of body unpatterned brown or dull red, or with indistinct black cross-bars; forehead brown; dark cross-bar crosses eye and reaches middle row of lip scales; dark forehead spot; a dark bar from forehead to sides of neck; lip scales pink with dark sutures; large chevron at back of head; belly yellowish-cream, with dark squarish marks. **DISTRIBUTION** Thailand (Tak, Uthai Thani and Chon Buri Provinces). Extralimitally: Laos and Cambodia. **HABITS AND HABITAT** Hill dipterocarp forests, up to 400m. Diet and reproductive habits unstudied.

Cambodian Kukri Snake ▪ *Oligodon mouhoti* 33.9cm

DESCRIPTION Top of body is brownish grey or greyish tan, the dorsals edged with dark brown; has a yellow, tan or light brown vertebral stripe, edged on each side with an ochre-brown paravertebral stripe; forehead is brownish grey or greyish brown; a broad, transverse reddish-brown mark on snout extends obliquely across eye; belly anterior is cream, rest of belly is bright pink-red or coral-red with irregular black blotches. Body is moderately built and sub-cylindrical; head is short and indistinct from neck; eyes are moderate in size with a rounded pupil; dorsals are smooth; subcaudals are paired; cloacal scute is entire. **DISTRIBUTION** Thailand. Extralimitally: Cambodia. **HABITS AND HABITAT** Lowland forests and near rice fields (altitude unknown). Terrestrial. Diet is unstudied. Oviparous, clutches comprising 2 eggs (26–28 × 9mm).

Eight-lined Kukri Snake ■ *Oligodon octolineatus* 70cm
(Bahasa Brunei: Ular Kalibantang. Bahasa Indonesia: Ular Birang. Iban: Ular Emparo, Ular Matahari)

DESCRIPTION Top of body is brown or reddish brown with 5–7 light (cream or white) longitudinal stripes, these bordered by 6–8 dark brown or black stripes; forehead has 2 black stripes; belly is yellow. Body is slender and sub-cylindrical; head is short and rounded, and is indistinct from neck; eyes are moderate in size with a rounded pupil; dorsals are smooth; cloacal scute is entire. **DISTRIBUTION** Thailand, Peninsular Malaysia, Singapore, Sumatra,

Pulau Nias, Mentawai and Riau archipelagos, Pulau Bangka, Pulau Belitung, Borneo, Java. Extralimitally: Sulu Archipelago, Sulawesi. **HABITS AND HABITAT** Lowland and hill dipterocarp forests (<500m asl). Mostly terrestrial, with occasional arboreal activity. Diet comprises bird eggs, frogs and their eggs, lizards and other snakes. Oviparous, producing clutches of 4–5 eggs (16 × 30mm).

Purple Kukri Snake ■ *Oligodon purpurascens* 95cm
(Bahasa Malaysia: Ular Kebun Perang. Iban: Ular Paut. Thai: Ngu Kut)

DESCRIPTION Top of body is brownish purple with dark, wavy bands or transverse dark-edged yellow bands or blotches that are separated by narrow, dark cross-bars; a dark chevron is present on forehead; belly is pink or red, with dark, squarish spots. Body is thick and sub-cylindrical; head is short and indistinct from neck; eyes are small with a rounded pupil; dorsals are smooth; cloacal scute is entire. **DISTRIBUTION** Thailand, Peninsular

Malaysia, Singapore, Sumatra, Mentawai Archipelago, Borneo, Java. **HABITS AND HABITAT** Hill dipterocarp forests and peat-swamp forests (<1,200m asl). Diet includes frog and lizard eggs, and frogs and tadpoles. Oviparous, clutches comprising 8–13 eggs (18–22 × 27–33mm).

Four-lined Kukri Snake ■ *Oligodon quadrilineatus* 40.2cm
(Thai: Ngu Ngod Lai Si Keet)

DESCRIPTION Top of body is greyish brown, scales narrowly edged with dark brown; a narrow greyish-yellow vertebral stripe, edged with a wide brown paravertebral stripe, runs from nuchal marking to tail base, and is intersected with small irregular, sub-rectangular, dark brown blotches; forehead is greyish brown; transverse dark brown mark on snout extends across eye to supralabials; has an arrow- or heart-shaped, dark brown nuchal mark; belly is cream-yellow, with 2 irregular sub-rectangular blackish-brown blotches near edges of ventrals. Body is robust and sub-cylindrical; head is short and indistinct from neck; eyes are moderate in size with a rounded pupil; dorsals are smooth; cloacal scute is entire. **DISTRIBUTION** Myanmar, Thailand. Extralimitally: Cambodia, Vietnam. **HABITS AND HABITAT** Lowland forests. Diet and reproductive habits are unstudied.

Half-keeled Kukri Snake ■ *Oligodon signatus* 39cm

DESCRIPTION Top of body is reddish brown with 20–30 dark-edged, light cross-bars; has a chevron pattern on forehead; belly is orange-red. Body is slender and sub-cylindrical; head is short and indistinct from neck; eyes are small with a rounded pupil; dorsals are weakly keeled; cloacal scute is entire. **DISTRIBUTION** Peninsular Malaysia, Singapore, Borneo. **HABITS AND HABITAT** Lowland forests (<200m asl). In buttresses of large trees and edges of peat-swamp forests. Diet and reproductive habits are unstudied.

Striped Kukri Snake ■ *Oligodon taeniatus* 44.7cm
(Thai: Ngu Ngod Thai)

DESCRIPTION Top of body is brownish grey or greyish tan, dorsals finely edged with dark brown posteriorly; has 2 dark longitudinal stripes edging a yellow vertebral stripe, and 2 narrower dorso-lateral stripes; 5 blotches are present on forehead; bases of oblique central streaks may reach ventrals; belly is pink or coral-red, cream on edges, with an irregular,

sub-rectangular black blotch near 1 or both tips. Body is moderately built and sub-cylindrical; head is short and indistinct from neck; eyes are small with a rounded pupil; dorsals are smooth; cloacal scute is entire. **DISTRIBUTION** Thailand. Extralimitally: Laos, Cambodia, Vietnam. **HABITS AND HABITAT** Lowland forests. Diet comprises frogs, lizards and their eggs. Oviparous (egg numbers and size unknown).

Mandalay Kukri Snake ■ *Oligodon theobaldi* 43.7cm
(Thai: Ngu Pi-kaeo Pama)

DESCRIPTION Top of body is light brown, with dark, narrow, close-set transverse or angular cross-bars; 4 dark longitudinal stripes run along dorsum; belly is yellow, with or without squarish black spots at outer margins of ventrals. Body is slender and sub-

cylindrical; head is short and indistinct from neck; eyes are moderate in size with a rounded pupil; tail is short; dorsals are smooth; cloacal scute is divided. **DISTRIBUTION** Myanmar. Extralimitally: E India. **HABITS AND HABITAT** Forested mid-hills and plains (altitude unknown). Diet is unstudied. Oviparous, clutches comprising 3 eggs (diameter 18mm).

Red Bamboo Rat Snake ■ *Oreocryptophis porphyraceus* 125cm
(Mandarin Chinese: P'an-tsu-hwa Zeh. Thai: Ngu Tang-ma-prao Daeng)

DESCRIPTION Top of body is deep reddish brown, with dark bands on body and tail; a dark flank stripe runs along posterior half of body; has a dark post-ocular stripe; belly is plain cream or yellow; iris is yellowish brown; tongue is reddish brown. Body is slender; head is elongated and slightly distinct from neck; snout is rounded; eyes are small with a rounded pupil; ventrals lack keels; cloacal scute is divided. **DISTRIBUTION** Myanmar, Thailand, Peninsular Malaysia, Sumatra. Extralimitally: E India, Bhutan; possibly also Nepal, China, Laos, Cambodia, Vietnam. **HABITS AND HABITAT** Forested mid-hills to montane forests (116–2,600m asl). In low vegetation. Diet comprises voles and shrews. Oviparous, laying clutches of 2–5 eggs (diameter 48mm).

Cave Racer ■ *Orthriophis taeniurus* 200cm
(Bahasa Malaysia: Ular Bulan. Mandarin Chinese: Heimeijing She. Iban: Ular Bulan. Kelabit: Rari Wang. Thai: Ngu Gab-mal Dam)

DESCRIPTION Top of body is greyish brown or greyish black with a cream or tan stripe along middle of back, especially towards posterior; forehead is olive; sides of head have a dark stripe; supralabials and chin are cream; belly is yellow or cream. Body is slender and elongate; head is long and distinct from neck; eyes are large with a rounded pupil; dorsals are weakly keeled; scales on flanks are smooth; cloacal scute is divided. **DISTRIBUTION** Myanmar, Thailand, Peninsular Malaysia, Singapore, Borneo, Sumatra. Extralimitally: Laos, Cambodia, Vietnam, E India, Bhutan, S China, Ryukyu Archipelago, SE Russia. **HABITS AND HABITAT** Lowland and sub-montane forests (<2,000m asl). Arboreal and terrestrial; often encountered in caves. Diet comprises bats and swiftlets. Oviparous, producing clutches of 5–14 eggs (45.7–72.0 × 20.8–31.4mm).

White-collared Reed Snake ■ *Pseudorabdion albonuchalis* 27cm

DESCRIPTION Top of body is iridescent black; a broad yellow or red collar covers more than half the parietals and a band of 4 scales behind parietals; belly is dark brown. Body

is slender; head is indistinct from neck; snout is pointed; has a single nostril, between 2 nasals; frontal borders eye; eyes are small with a vertical pupil; tail is short; dorsals are smooth; cloacal scale is entire. **DISTRIBUTION** Borneo. **HABITS AND HABITAT** Lowland dipterocarp forests (<500m asl). Sub-fossorial, in leaf litter. Diet is unstudied but presumably includes small arthropods. Reproductive habits are unstudied.

Dwarf Reed Snake ■ *Pseudorabdion longiceps* 23cm
(Iban: Ular Untup. Thai: Ngu Pong-oa Hua Yao)

DESCRIPTION Top of body is iridescent black or brown; a yellow collar and yellow spot are present above angle of mouth; belly is dark brown. Body is slender; head is indistinct from neck; snout is pointed; nostril is situated within a single nasal; eyes are small with a rounded pupil; tail is short; dorsals are smooth; caudal scale is entire. **DISTRIBUTION**

Thailand, Peninsular Malaysia, Singapore, Sumatra, Pulau Nias, Mentawai and Riau archipelagos, Borneo. Extralimitally: Sulawesi. **HABITS AND HABITAT** Lowland rainforests (<500m asl). Sub-fossorial. Diet comprises litter invertebrates, earthworms, and small insects and larvae. Oviparous, laying clutches of 2–3 narrow, elongate eggs (size unknown).

Sarawak Reed Snake ■ *Pseudorabdion saravacensis* 142mm

DESCRIPTION Top of body is dark brown, iridescent; red blotches present on sides of head above angle of jaws; irregular red band present on neck; belly is unpatterned brown. Body slender; head indistinct from neck; snout pointed; eyes small; tail short and pointed; dorsals smooth; cloacal scute entire. **DISTRIBUTION** Borneo. **HABITS AND HABITAT** Lowland forests. Nothing known of its natural history.

Keeled Rat Snake ■ *Ptyas carinata* 400cm
(Thai: Ngu Sing Hang Dam)

DESCRIPTION Top of body is olive-brown to nearly black anteriorly, sometimes with indistinct yellow cross-bars; posterior dorsum is yellow with a distinct chequered black pattern, ending in a black tail with yellow spots; belly is cream, turning grey or black posteriorly. Body is slender; head is distinct from neck; eyes are large with a rounded pupil; tail is long; dorsals are smooth, except for the 2–4 median rows that are keeled; cloacal scute is divided.

DISTRIBUTION Myanmar, Thailand, Peninsular Malaysia, Singapore, Sumatra, Borneo, Java. Extralimitally: Vietnam, S China, Palawan. **HABITS AND HABITAT** Lowland forests and agricultural fields (altitude unknown). Diet comprises amphibians and rodents. Oviparous, clutches comprising 10 eggs (size unknown).

White-bellied Rat Snake ▪ *Ptyas fusca* 300cm
(Thai: Ngu Sing Tai)

DESCRIPTION Top of body is mid-brown to brownish grey, typically nearly black anteriorly and lighter posteriorly; a red vertebral stripe is sometimes present; belly is cream. Body is slender; head is distinct from neck; eyes are large with a rounded pupil; tail is long; dorsals are smooth; cloacal scute is divided. **DISTRIBUTION** Peninsular Malaysia, Singapore, Sumatra, Pulau Nias, Pulau Bangka, Pulau Belitung, Natuna Archipelago, Borneo. **HABITS AND HABITAT** Lowland forests to mid-hills (<1,330m asl). Diet includes birds and, probably, small mammals. Oviparous (egg numbers and size unknown).

Javanese Rat Snake ▪ *Ptyas korros* 268cm
(Bahasa Malaysia: Ular Tikus Biasa. Bahasa Indonesia: Ular Koros. Sundanese: Oraj Lingas. Thai: Ngu Sing Ban)

DESCRIPTION Top of body anterior and forehead grey to olive-brown, body darkening to nearly black posteriorly; scales edged with white, becoming more distinct posteriorly, where they appear as white bands on a black background; belly, chin and labials are brownish cream. Body is robust; head is elongate and distinct from neck; eyes are large

with a rounded pupil; tail is long; dorsals are smooth anteriorly, weakly keeled posteriorly; cloacal scute is divided. **DISTRIBUTION** Myanmar, Thailand, Peninsular Malaysia, Singapore, Sumatra, Borneo, Java. Extralimitally: Laos, Cambodia, Vietnam, E India, Bhutan, Bangladesh, E China. **HABITS AND HABITAT** Lowland and montane forests (<3,000m asl). Terrestrial, but can climb trees to rest and to mate. Diet includes rodents, birds, lizards and frogs. Oviparous, clutches comprising 4–14 eggs (33.5–48 × 17.1–20.5mm).

Indian Rat Snake ■ *Ptyas mucosa* 370cm
(Bahasa Indonesia: Ular Jali Belang. Thai: Ngu Sing Hang Lai)

DESCRIPTION In adults, top of body ranges from yellowish brown and olivaceous brown to black; posterior of body has dark bands or a reticulate pattern; belly is greyish white or yellow. Hatchlings are greenish brown, typically with light bluish-grey cross-bands on anterior of body. Body is slender; head is elongate and distinct from neck; eyes are large with a rounded pupil; tail is long; dorsals are smooth or weakly keeled; cloacal scute is divided. **DISTRIBUTION** Myanmar, Thailand, Peninsular Malaysia, Singapore, Sumatra. Extralimitally: Laos, Cambodia, Vietnam, Turkmenistan, Iran, Pakistan, India, Nepal, Sri Lanka, Bhutan, Bangladesh, S China. **HABITS AND HABITAT** Lowland forests, agricultural fields, parks and gardens (<4,000m asl). Terrestrial. Diet comprises rats, frogs, bats, birds, lizards, turtles and snakes. Oviparous, clutches comprising 5–25 eggs (42–69 × 22.5–28.5mm).

Green Rat Snake ■ *Ptyas nigromarginata* 191.5cm

DESCRIPTION Top of body is a soft velvety green or olive-green; dorsals are black-edged; 4 longitudinal black stripes run along body and tail in juveniles, and in adults are confined to posterior third of body; head is olive-brown, with a bright yellow patch on temporal region; belly has a greenish-yellow tinge. Body is slender and slightly compressed; head is elongate; eyes are large with a rounded pupil; tail is long; dorsals are smooth, 4–6 median scale rows distinctly keeled; cloacal scute is divided. **DISTRIBUTION** Myanmar and Thailand. Extralimitally: Vietnam, E India, Bhutan, Nepal, S China. **HABITS AND HABITAT** Disturbed habitats and open forests (<720m asl). Terrestrial as well as arboreal. Diet comprises rodents, birds, lizards and snakes. Oviparous, clutches comprising 8–9 eggs (size unknown).

Collared Black-headed Snake ■ *Sibynophis collaris* 76cm
(Thai: Ngu Ko-kwan Dam)

DESCRIPTION Top of body is brown or greyish brown; has a black vertebral stripe comprising black spots, and occasionally a light, dotted dorso-lateral line; has a dark cross-bar behind eye and another on forehead; has a black transverse band on nape, bordered posteriorly with yellow; belly is yellow, ventrals with black spot on outer edges, forming a dark, dotted line on either side. Body is slender and cylindrical; head is relatively short and flattened, and slightly distinct from neck; eyes are large with a rounded pupil; dorsals

are smooth; cloacal scute is divided. **DISTRIBUTION** Myanmar, Thailand, Peninsular Malaysia. Extralimitally: Laos, Cambodia, Vietnam, India, Bangladesh, Nepal. **HABITS AND HABITAT** Forests, from lowland plains to montane limits (<3,050m asl). Terrestrial. Diet comprises skinks, snakes, frogs and insects. Oviparous, laying 4–6 eggs (28 × 15mm).

White-lipped Black-headed Snake ■ *Sibynophis melanocephalus* 60cm
(Thai: Ngu Ko-kwan Hua Dam)

DESCRIPTION Top of body is reddish brown or brown, with short black cross-bars over a lighter band; lower flanks have small yellow spots; forehead is dark olive with olive-yellow spots; supralabials are white, sutures with black stripes; belly is yellow with an orange

tinge on sides posteriorly; ventrals have a rounded black spot laterally. Body is slender and cylindrical; head is relatively short, slightly distinct from neck and flattened; eyes are large with a rounded pupil; dorsals are smooth; cloacal scute is divided. **DISTRIBUTION** Thailand, Peninsular Malaysia, Singapore, Sumatra, Borneo. Extralimitally: Vietnam. **HABITS AND HABITAT** Lowland forests (<500m asl). Terrestrial, on leaf litter on the forest floor and by streams. Exhibits caudal autotomy. Diet comprises skinks. Reproductive habits are unstudied.

Two-coloured Forest Snake ■ *Smithophis bicolor* 60cm

DESCRIPTION Top of body is dark brown or black, clearly separated from yellowish-cream belly; subcaudals are cream, unpatterned or with black spots. Body is robust, cylindrical and elongate; head is indistinct from neck and depressed; snout is rounded; nostril is dorso-laterally located; eyes are small with rounded or elliptical pupil; dorsals are smooth; cloacal scute is divided. **DISTRIBUTION** Myanmar. Extralimitally: E India, S China. **HABITS AND HABITAT** Forested mid-hills (altitude unknown). Terrestrial or sub-fossorial. Diet includes earthworms and slugs. Reproductive habits are unstudied.

Bornean Black Snake ■ *Stegonotus borneensis* 137cm
(Thai: Ngu Daeng Borneo)

DESCRIPTION Top of body is plain grey-black, dark brown or black; supralabials are grey with a pink tinge; each scale on belly is alternately banded with dark and light grey. Body is robust and cylindrical; head is distinct from neck; a distinct vertebral ridge is present; eyes are small with a vertical pupil; dorsals are smooth; vertebrals are enlarged; cloacal scute is entire. **DISTRIBUTION** Borneo, Thailand. **HABITS AND HABITAT** Low hills to sub-montane limits (<1,800m asl). Terrestrial. Diet is unstudied. Oviparous (egg numbers and size unknown).

Fruhstorfer's Mountain Snake ■ *Tetralepis fruhstorferi* 50cm

DESCRIPTION Back of body dark reddish-brown; black vertebral stripe; belly bluish- or reddish-grey, each ventral scale with two dark brown spots that form two longitudinal rows of stripes. **DISTRIBUTION** Java (Gunung Tengger). **HABITS AND HABITAT** Submontane forests, at elevations of 1,200m asl. Diet and reproductive habits unstudied. NOTE Familial placement is tentative.

Ornate Brown Snake ■ *Xenelaphis ellipsifer* 251cm

DESCRIPTION Top of body is mid-brown, with 18–20 large elliptical or squarish, black-edged brown blotches (1–2 scale rows wide), separated by narrow cream interspaces; forehead is plain reddish brown; supralabials are yellow with black markings; infralabials are plain cream; neck has dark longitudinal streaks; flanks have inverted Y- or V- shaped markings; belly is pink or white, with dark spots on outer edges of ventrals. Body is robust; head is distinct from neck; snout is rounded; eyes are large with a rounded pupil; tail is long; dorsals are smooth; cloacal scute is divided. **DISTRIBUTION** Peninsular Malaysia, Sumatra, Borneo. **HABITS AND HABITAT** Primary forests (800–1,000m asl). Aquatic. Diet unknown but presumably comprises fish. Reproductive habits are unstudied.

Malayan Brown Snake ■ *Xenelaphis hexagonotus* 200cm
(Iban: Ular Beluai. Thai: Ngu Kuan Kanun)

DESCRIPTION In adults, top of body is brown, dark green or greenish olive, with narrow, vertical black bars extending from neck and along body, the apices reaching lower flanks; belly is pale or deep yellow, ventrals with a dark spot on margins. In juveniles, top of body is uniformly light brown. Body is robust; head is distinct from neck; snout is rounded; eyes are large with a rounded pupil; tail is long; vertebrals are enlarged and hexagonal in shape; dorsals are smooth; cloacal scute is divided. **DISTRIBUTION** Myanmar, Thailand, Peninsular Malaysia, Singapore, Sumatra, Pulau Bangka, Pulau Belitung, Borneo, Java. Extralimitally: Vietnam. **HABITS AND HABITAT** Lowland forests, especially coastal peat swamps and mangrove swamps (<500m asl). Terrestrial, in waterlogged forests. Diet comprises rodents, frog and fish. Reproductive habits are unstudied.

> **LAMPROHIIDAE – AFRO-EURASIAN SNAKES**
> Members of this recently recognised snake family are Afro-Eurasian in distribution.
> Most members are terrestrial, oviparous and feed on small vertebrates. They are non-
> venomous and tend to inhabit more open landscapes.

Indo-Chinese Sand Snake ■ *Psammophis indochinensis* 107.5cm
(Bahasa Indonesia: Ular Pasir Asia. Thai: Ngu Man-tong)

DESCRIPTION Top of body is olive-green or buff; interstitial region is black, with 4 dark brown stripes (2 scales broad) extending from brown forehead to tail; belly is bright yellow or yellowish cream, with a black line at outer margin of ventrals. Body is slender; head is oval and distinct from neck; eyes are large with a rounded pupil; dorsals are smooth; cloacal scute is divided. **DISTRIBUTION** Myanmar, Thailand, Java, Bali. Extralimitally: Laos, Vietnam. **HABITS AND HABITAT** Grasslands, open forests and agricultural areas (100–2,000m asl). Terrestrial, with some arboreal activity in bushes and short trees. Diet comprises rodents, frogs, lizards and snakes. Oviparous (egg numbers and size unknown).

NATRICIDAE – WATER SNAKES
This family includes the large water snakes of the Old and New worlds, which lack enlarged grooved rear fangs on the maxillary bone. Their habitats range from terrestrial to highly aquatic, and from mountain-tops to bodies of fresh water, where they feed on fish and other aquatic animals. Cosmopolitan in distribution, they may be found in both tropical and temperate regions of Asia, Europe, Africa, Australia and South America.

Buff-striped Keelback ▪ *Amphiesma stolatum* 80cm
(Thai: Ngu Lai-sab Dok-ya)

DESCRIPTION Top of body is olive-grey to greenish grey; pattern is variable depending on locality, but typically includes buff, orange, pale yellow or orangish-yellow dorso-lateral stripes on 5th–7th scale rows; belly is pale. Body is robust; head is distinct from neck; eyes are large with a rounded pupil; tail is short; dorsals are keeled; cloacal scute is divided. **DISTRIBUTION** Myanmar, Thailand. Extralimitally: Laos, Cambodia, Vietnam, India, Bhutan, Pakistan, Nepal, Sri Lanka, Bangladesh, China. **HABITS AND HABITAT** Grasslands and lowland forests (<2,000m asl). Terrestrial, in grassy areas, especially near lakes, ponds and rice fields. Diet comprises insects, frogs, scorpions, fish and lizards. Oviparous, laying 3–15 eggs (22–35 × 12–18mm).

Two-lined Keelback ■ *Hebius bitaeniatum* 70.8cm

DESCRIPTION Top of body is dark brownish grey or ochre-brown, with a broad beige-yellow dorso-lateral stripe, edged by 2 narrow black lines, running from neck to tail tip; forehead is greyish brown with indistinct vermiculation; supralabials are ivory-yellow; a

dark brown post-ocular stripe extends across neck to join dorso-lateral stripe; belly is yellowish cream. Body is moderately built and elongate; head is distinct from neck; nostrils are lateral; eyes are large with a rounded pupil; tail is long and tapering; dorsals are keeled, the scales being notched posteriorly; subcaudals are paired; cloacal scute is divided. **DISTRIBUTION** Myanmar, Thailand. Extralimitally: Laos, Vietnam, S China. **HABITS AND HABITAT** Montane forests (1,829–2,377m asl). Terrestrial. Diet and reproductive habits are unstudied.

Boulenger's Keelback ■ *Hebius boulengeri* 87.7cm

DESCRIPTION Top of body is greyish black or brown, with a pair of longitudinal white dorsal stripes on head that turn pinkish brown posteriorly; forehead is brownish grey with fine grey vermiculation; a distinct narrow white post-ocular streak extends from lower margin of orbit to nape; belly is cream, with square blotches on edges of ventrals. Body is

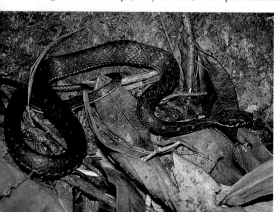

slender; head is distinct from neck; nostrils are lateral; eyes are large with a rounded pupil; dorsals are weakly keeled, except outermost rows, which are smooth; subcaudals are paired; cloacal scute is divided. **DISTRIBUTION** Thailand. Extralimitally: Laos, Cambodia, Vietnam, E China. **HABITS AND HABITAT** Temperate forests in mid-hills (540–900m asl). Terrestrial, on the forest floor and near rocky streams. Diet is unstudied. Oviparous, laying 3 eggs (22–29 × 7–11mm).

Deschauensee's Keelback ▪ *Hebius deschauenseei* 48cm
(Thai: Ngu Lai-sab Tong Sam Keet)

DESCRIPTION Top of body is brownish yellow; flanks have irregular black blotches; forehead and neck are black, this coloration extending dorsally as a vertebral stripe; belly is

pale, turning darker posteriorly, with a series of 3 grey or brown spots. Body is slender; head is distinct from neck; nostrils are dorso-lateral; eyes are small with a rounded pupil; dorsals are keeled; cloacal scute is divided. **DISTRIBUTION** Thailand. Extralimitally: Vietnam, China. **HABITS AND HABITAT** Lowland forests (*c.* 300m asl). Terrestrial, near forest streams. Diet and reproductive habits are unstudied.

White-fronted Keelback ▪ *Hebius flavifrons* 75cm
(Bahasa Malaysia/Indonesia: Ular Hidung Putih. Iban: Ular Ginti)

DESCRIPTION In adults, top of body is olive-grey with darker markings; has a distinctive

white to yellowish-cream spot on snout; belly is cream with an alternating series of black spots. Juveniles have paired white spots along mid-back. Body is slender; head is distinct from neck; eyes are large with a rounded pupil; dorsals are keeled; cloacal scute is divided. **DISTRIBUTION** Borneo. **HABITS AND HABITAT** Lowlands, in foothills. Aquatic, in rivers. Diet includes frogs, their eggs and tadpoles. Reproductive biology is unstudied.

Gunung Inas Keelback ■ *Hebius inas* 61.5cm
(Thai: Ngu Lai-sab Malayu)

DESCRIPTION Top of body is dark olive-brown, with rows of indistinct dorso-lateral spots that fuse to form a pale stripe extending posteriorly to meet transverse bars on dorsum;

flanks have faint yellow spots; forehead is brown, variegated with black; posterior supralabials have dark, rounded blotches; chin and throat have black spots; belly is white with a squarish black spot at outer margin of ventrals. Body is slender; head is distinct from neck; eyes are large with a rounded pupil; dorsals are distinctly keeled; cloacal scute is divided. **DISTRIBUTION** Thailand, Peninsular Malaysia. **HABITS AND HABITAT** Mid-hill forests. Terrestrial, active on forest litter and tree buttresses. Diet and reproductive habits are unstudied.

Khasi Hills Keelback ■ *Hebius khasiense* 60cm

DESCRIPTION Top of body has 3 blackish-brown stripes, alternating with 4 reddish-brown stripes; forehead is greyish brown with short, pale brown marks; labials are white with dark

edges, this marking continuing to sides of neck; posterior supralabials have rounded blotches; belly is white; ventrals have reddish-brown outer edges. Body is slender; head is distinct from neck; eyes are large with a rounded pupil; dorsals are keeled; cloacal scute is divided. **DISTRIBUTION** Myanmar, Thailand. Extralimitally: Laos, Vietnam, S China, E India. **HABITS AND HABITAT** Sub-montane forests (900–1,000m asl). Terrestrial, on the forest floor near streams. Diet comprises insects, frogs and tadpoles. Oviparous (numbers and size of eggs unknown).

Striped Keelback ▪ *Hebius parallelum* 63.5cm

DESCRIPTION Top of body is olive-brown or greyish brown, with 2 light dorso-lateral stripes or a series of spots extending from neck to tail tip; forehead is brown; has a short yellow vertebral streak behind occiput and a black streak from eye to angle of mouth; belly is plain yellow with a row of black dots on each side. Body is robust; head is distinct from neck; eyes are large with a rounded pupil; dorsals are strongly keeled, except for outer row; cloacal scute is divided. **DISTRIBUTION** Myanmar. Extralimitally: Nepal, E India, S China. **HABITS AND HABITAT** Sub-montane forests (c. 1,000m asl). Diet comprises fish. Oviparous (numbers and size of eggs unknown).

Red Mountain Keelback ■ *Hebius sanguineum* 60cm

DESCRIPTION Top of body is crimson or orangish yellow, with a vertebral band comprising 4–5 rows of elongated oval olive and diamond-shaped black marks; flanks

have 2 alternating rows of black spots; forehead is dark olive; a black-edged white post-ocular stripe extends to nape; labials are cream with black sutures; belly is crimson with an indistinct black spot on outer edges of ventrals; chin and throat are plain cream. Body is slender; head is distinct from neck; eyes are large with a rounded pupil; dorsals are keeled; cloacal scute is divided. **DISTRIBUTION** Peninsular Malaysia. **HABITS AND HABITAT** Montane forests of Cameron Highlands, Pahang, and Gombak Valley, Selangor. Terrestrial. Diet and reproductive habits are unstudied.

Sarawak Keelback ■ *Hebius saravacense* 78cm

DESCRIPTION Top of body is olive to reddish brown, dorsum with squarish black markings; has a row of light spots on flanks; supralabials are yellow or cream; belly is yellow and black. Body is slender; head is distinct from neck; eyes are large with a rounded pupil; tail is long; dorsals are keeled; cloacal scute is divided. **DISTRIBUTION** Peninsular Malaysia, Borneo. **HABITS AND HABITAT** Mid-hills and sub-montane forests (640–1,700m asl). Terrestrial, near streams. Diet comprises frogs and their eggs. Oviparous, producing clutches of 4–5 eggs (size unknown).

Venning's Keelback ■ *Hebius venningi* 78cm

DESCRIPTION Top of body is olive-brown, indistinctly chequered with black and anteriorly marked with dorso-lateral ochre spots; scales are flecked with black; forehead has lighter vermicular marks; supralabials are pale ochre with black posterior margins; belly is coral-red, with fawn outer edges smudged with black. Body is slender; head is distinct from neck; eyes are large with a rounded pupil; inner dorsals are weakly keeled, outer dorsals are smooth; upper rows are weakly keeled; cloacal scute is divided. **DISTRIBUTION** Myanmar. Extralimitally: E India, Bangladesh, S China. **HABITS AND HABITAT** Evergreen forests (1,040–1,400m asl). Terrestrial, on the forest floor near hill streams. Diet includes tadpoles and frogs. Reproductive habits are unstudied.

Strange-tailed Keelback ■ *Hebius xenura* 66cm

DESCRIPTION Top of body is olive-brown to nearly black, with a paired series of reddish-orange, pale brown, yellow or white spots on flanks; adjacent spots may be connected by faint, dark cross-lines; labials are white, with dark lines on sutures; belly is white or yellow, outer edges of ventrals with dark brown spots. Body is slender; head is distinct from neck; eyes are large with a rounded pupil; dorsals are keeled; cloacal scute is entire or divided. **DISTRIBUTION** Myanmar. Extralimitally: E India, Bangladesh. **HABITS AND HABITAT** Sub-montane forests. Terrestrial, on the forest floor near streams. Diet and reproductive habits are unstudied.

Siebold's Keelback ■ *Herpetoreas sieboldii* 94.3cm

DESCRIPTION Top of body is plain olive-green or brown; has a series of small white dorso-lateral spots; forehead is brown, paler on flanks; has a pair of pale occipital spots and a post-parietal streak; supralabials are bordered dorsally by a dark stripe that reaches

the nuchal region as a crescent; belly is light, typically with dark greyish-brown speckles posteriorly. Body is slender; head is distinct from neck; eyes are large with a rounded pupil; dorsals are keeled; cloacal scute is divided. **DISTRIBUTION** Myanmar. Extralimitally: Bhutan, India, Nepal, Pakistan. **HABITS AND HABITAT** Sub-montane and montane forests (1,219–3,658m asl). Terrestrial. Diet comprises frogs and their eggs, tadpoles and skinks. Oviparous, clutches comprise 5 eggs (size unknown).

Yellow-spotted Water Snake ■ *Hydrablabes periops* 53cm

DESCRIPTION Top of body is olive-brown, sometimes with a pale brown stripe on flanks; some individuals are unpatterned; belly is yellow or grey. Body is slender; head is small and distinct from neck; eyes are small with a rounded pupil; tail is short; dorsals are smooth; cloacal scute is divided. **DISTRIBUTION** Borneo. **HABITS AND HABITAT** Lowland and mid-hill dipterocarp forests (150–600m asl). Aquatic, near streams. Diet and reproductive habits are unstudied.

Corrugated Water Snake ■ *Opisthotropis typica* 50.2cm

DESCRIPTION Top of body is plain blackish grey; lip scales are grey; belly is plain cream. Body is slender; head is small, depressed and indistinct from neck; forehead scales are finely striated; eyes are small with a rounded pupil, and separated from labials by small scales; dorsals are strongly keeled; cloacal scute is divided. **DISTRIBUTION** Borneo. Extralimitally: Palawan. **HABITS AND HABITAT** Lowland dipterocarp forests (75–500m asl). Aquatic, in shallow rocky streams and swampy pools. Diet comprises tadpoles. Reproductive biology is unstudied.

Brown Stream Snake ■ *Paratapinophis praemaxillaris* 98cm
(Thai: Ngu Lai-so Lao Noe)

DESCRIPTION Top of body is plain brownish grey; edges of dorsals and ventrals are greyish brown; anterior dorsum has an indistinct yellow inverted V-shaped mark, which in females has a dark blotch within it and turns into blue-grey bands extending across top (in males, bluish-grey colours are absent); forehead is plain brownish grey; supralabials are brownish grey; belly is cream. Body is robust, quadrangular in cross section in females and relatively slender in males; head is distinct from neck; eyes are large and lateral with a rounded pupil; dorsals are smooth, scales with middle row of tubercles; subcaudals are paired; cloacal scute is divided. **DISTRIBUTION** Thailand. Extralimitally: Laos, China. **HABITS AND HABITAT** Subtropical alpine forests (475–1,400m asl). Aquatic, in fast-flowing rocky streams. Diet comprises fish. Oviparous (egg numbers and size unknown).

Painted Mock Viper ■ *Psammodynastes pictus* 55cm
(Bahasa Malaysia: Ular Sampah. Bahasa Indonesia: Ular Berang, Ular Percha)

DESCRIPTION Top of body is brown, tan or black, with dark-edged light transverse bands; a dark streak is present along eyes; belly is cream with brown speckles. Body is slender; head is flattened and distinct from neck; eyes are large with a vertical pupil; dorsals are smooth; subcaudals are paired; cloacal scute is entire. **DISTRIBUTION** Peninsular Malaysia, Singapore, Sumatra, Riau and Mentawai archipelagos, Pulau Belitung, Borneo. **HABITS AND HABITAT** Lowland forests. Diet comprises fish, frog, lizards and prawns. Ovoviviparous (neonate numbers and size unknown).

Mock Viper ■ *Psammodynastes pulverulentus* 77cm
(Bahasa Indonesia: Ular Viper Tiruan. Thai: Ngu Mok)

DESCRIPTION Top of body is reddish brown to yellowish grey and nearly black, with small dark spots or streaks; usually, a longitudinal stripe is present along mid-dorsal region and 3 longitudinal stripes along flanks; belly is spotted with brown or grey, and with dark spots or longitudinal lines. Body is slender; head is flattened and distinct from neck; snout is short; eyes are large with a vertical pupil; dorsals are smooth, lacking apical pits; subcaudals are paired; cloacal scute is divided. **DISTRIBUTION** Myanmar; Thailand; Peninsular Malaysia; Sumatra; Pulau Bangka; Mentawai, Natuna and Riau archipelagos; Borneo; Java. Extralimitally: Laos, Cambodia, Vietnam, E India, Bhutan, Nepal, S China, Sulawesi, Lesser Sundas, Philippines, Taiwan. **HABITS AND HABITAT** Evergreen and tropical forests (<2,000m asl). Terrestrial and semi-arboreal. Diet comprises skinks, snakes, frogs and geckos. Ovoviviparous, producing 3–10 neonates (148–178mm).

Speckle-bellied Keelback ■ *Rhabdophis chrysargos* 98cm
(Bahasa Malaysia: Ular Rabong Perut Bintik. Bahasa Indonesia: Ular Kadut. Thai: Ngu Lai-sab Chut Dam Kao)

DESCRIPTION Top of body is olive-grey or olive-brown; supralabials are yellow or cream with darker smudges; a cream nuchal chevron, edged with black, is present; a reddish-brown or orange band runs behind neck; rest of back has yellow and brown oblong marks within darker bands; belly is yellow with brown mottling. Body is moderate and cylindrical; head is distinct from neck; eyes are large with a rounded pupil; tail is short; dorsals are keeled; subcaudals are paired; cloacal scute is divided. **DISTRIBUTION** Myanmar, Thailand, Peninsular Malaysia, Sumatra, Pulau Nias, Pulau Simeulue, Mentawai and Anamba archipelagos, Borneo, Java, Bali. Extralimitally: Laos, Cambodia, Vietnam, Sulawesi, Flores, Ternate, Palawan, Balabac. **HABITS AND HABITAT** Evergreen and tropical sub-montane forests (100–1,676m asl). Terrestrial, near streams. Diet includes frogs. Oviparous, laying 3–10 eggs (12–21 × 19.5–34mm). **VENOM** Unknown, with insufficient clinical reports; no antivenoms available; those raised for related species may be used for significant envenoming.

Red-bellied Keelback ■ *Rhabdophis conspicillatus* 55cm

DESCRIPTION Top of body is brown to reddish brown; sides of head have a downward-curving cream post-ocular stripe; supralabials are cream; nape and neck have 2 narrow cream collars; belly is yellow, each ventral dark-edged. Body is moderately built and cylindrical; head is distinct from neck; eyes are large with a rounded pupil; tail is short; dorsals are keeled, except smooth outer rows; subcaudals are paired; cloacal scute is divided. **DISTRIBUTION** Peninsular Malaysia, Sumatra, Pulau Singkep, Natuna Archipelago, Borneo. **HABITS AND HABITAT** Lowland and mid-hill forests (100–1,000m asl. Terrestrial, near water bodies, in leaf litter, tree buttresses or rotting logs, and under stones. Diet and reproductive habits are unstudied. **VENOM** No venom glands present.

Orange-lipped Keelback ■ *Rhabdophis flaviceps* 85cm
(Iban: Ular Kentu. Thai: Ngu Rang-hae Hua Daeng)

DESCRIPTION Top of body is greyish black with faint, light cross-bars, narrow at vertebral region and wide on flanks; forehead is light brown, yellowish brown or olive; supralabials

are rusty orange; a black-edged orange nuchal loop is present, especially in juveniles; belly is black or dark green, with black bands. Body is robust; head is distinct from neck; eyes are large with a rounded pupil; tail is short; dorsals are keeled; a nuchal gland is present on nape; cloacal scute is divided. **DISTRIBUTION** Thailand, Peninsular Malaysia, Sumatra, Pulau Bangka, Pulau Nias, Borneo. **HABITS AND HABITAT** Lowland and sub-montane forests (<1,300m asl). Semi-aquatic. Diet includes frogs, toads, tadpoles and lizards. Reproductive biology is unstudied.

Himalayan Keelback ■ *Rhabdophis himalayanus* 125cm

DESCRIPTION Top of body is olive to olive-brown or dark brown; 2 dorso-lateral rows of orange-yellow spots are present; anterior of body is chequered with vermilion spots; neck has a black-edged, bright yellow collar; labials are yellow, edged with black; a black sub-ocular stripe is present; belly is yellowish white. Body is moderately built and

cylindrical; head is distinct from neck; eyes are large with a rounded pupil; dorsals are keeled; subcaudals are paired; cloacal scute is divided. **DISTRIBUTION** Myanmar. Extralimitally: E India, Nepal, Bhutan, Bangladesh, S China. **HABITS AND HABITAT** Mid-hill and sub-montane forests (<1,100m asl). Terrestrial, on rocky biotopes and agricultural fields, near streams. Diet comprises frogs, lizards and fish. Oviparous, laying 5–7 eggs (size unknown). **VENOM** Unknown, with insufficient clinical reports; no antivenoms available; those raised for related species may be used for significant envenoming.

Gunung Murud Keelback ▪ *Rhabdophis murudensis* 87.3cm

DESCRIPTION Top of body is brownish grey with indistinct, dark cross-bars, these with a row of light spots on their edges; supralabials are bright red or brown; belly is greyish yellow with small black spots. Body is moderately built and cylindrical; head is distinct from neck; eyes are large with a rounded pupil; dorsals are keeled, except smooth outer rows; subcaudals are paired; cloacal scute is divided. **DISTRIBUTION** Gunung Murud and Gunung Mulu, Sarawak, and Gunung Kinabalu and Trus Madi, Sabah, Borneo. **HABITS AND HABITAT** Mid-hills, sub-

montane and montane forests (915–2,500m asl). Diet includes frogs. Reproductive biology is unstudied. **VENOM** Unknown, with insufficient clinical reports; no antivenoms available; those raised for related species may be used for significant envenoming.

Black-banded Keelback ▪ *Rhabdophis nigrocinctus* 95cm
(Thai: Ngu Lai-sab Kieo Kwan Dam)

DESCRIPTION Top of body is olive-green, turning more brown posteriorly, with indistinct, narrow black cross-bars; 2 black oblique stripes are present on flanks; forehead is copper-brown, paler on sides; has oblique black sub-ocular and post-ocular stripes; another stripe is present on nape; belly is white; subcaudals are pale pink, mottled with dark grey. Body is moderately built and cylindrical; head is distinct from neck; eyes are large with a rounded pupil; dorsals are keeled; subcaudals are paired; cloacal scute is divided. **DISTRIBUTION** Myanmar, Thailand. Extralimitally: Laos, Cambodia, Vietnam, S China. **HABITS AND HABITAT** Evergreen forests. Terrestrial and aquatic. Diet includes fish and frogs. Reproductive habits are unstudied. **VENOM** Unknown, with insufficient clinical reports; no antivenoms available; those raised for related species may be used for significant envenoming.

Collared Keelback ■ *Rhabdophis nuchalis* 62cm

DESCRIPTION Top of body is light brown, chequered with pale reddish-brown spots; forehead is brown, speckled with red; 2 oblique black stripes are present between posterior supralabials; rows of body scales alternate red and brown; interstitial area is bluish black; oblique black lines are present on dorso-lateral scales at mid-body; in adults, neck is reddish brown from angle of jaw posteriorly for 11–12 scale rows; juveniles have a bright reddish-yellow collar, posteriorly with a reddish tinge. Body is moderately built and cylindrical; head

is flat and distinct from neck; nuchal groove is distinct; eyes are large with a rounded pupil; dorsals are keeled, except smooth outer row; subcaudals are paired; cloacal scute is divided. **DISTRIBUTION** Myanmar. Extralimitally: Vietnam, E India, China. **HABITS AND HABITAT** Subtropical broadleaf forests and terrace cultivation (1,500–1,750m asl). Terrestrial. Diet is unstudied. Oviparous, producing 8–19 eggs (26 × 13mm). **VENOM** Unknown, with insufficient clinical reports; no antivenoms available; those raised for related species may be used for significant envenoming.

Olive Keelback ■ *Rhabdophis plumbicolor* 48.5cm

DESCRIPTION In adults, top of body is grass-green; juveniles have a large V-shaped mark on neck, followed by a similar smaller one, the intervening area yellow or orange; a black post-ocular stripe runs to angle of jaws; black spots or cross-bars are present on dorsum

and tail; belly is white, cream or grey. Body is robust; head is short and broad, distinct from neck; eyes are large with a rounded pupil; tail is short; dorsals are strongly keeled; a nuchal gland is present on nape; cloacal scute is divided. **DISTRIBUTION** Myanmar. Extralimitally: India, Sri Lanka. **HABITS AND HABITAT** Low hills (<2,134m asl). Terrestrial, in grassy habitats. Diet comprises toads. Reproductive habits are unstudied.

Blue-necked Keelback ■ *Rhabdophis rhodomelas* 75cm
(Thai: Ngu Rang-hae Lang Son)

DESCRIPTION Top of body is reddish brown; a black vertebral stripe enters nape as an inverted chevron; nape is light blue posteriorly; belly is pink, each ventral has a small dark spot. Body is robust; head is distinct from neck; eyes are large with a rounded pupil; tail is short; dorsals are keeled; a nuchal gland is present on nape; subcaudals are paired; cloacal scute is divided. **DISTRIBUTION** Thailand, Peninsular Malaysia, Singapore, Sumatra, Borneo, Java. **HABITS AND HABITAT** Streams and other wetlands in lowlands and mid-hills (<1,000m asl). Semi-aquatic. Diet comprises frogs, toads and tadpoles. Oviparous, laying 25 eggs (size unknown).

Red-necked Keelback ■ *Rhabdophis subminiatus* 130cm
(Bahasa Indonesia: Ular Picung. Thai: Ngu Lai-sab Kor Daeng)

DESCRIPTION In adults, top of body is olive-brown or green, plain or with black and yellow reticulation; nape has a yellow and red band; a dark oblique sub-ocular bar is present; belly is yellow, sometimes with a black dot on outer edge of ventrals. Juveniles have a black cross-bar or triangular mark on nape, bordered with yellow posteriorly; dorsum has oval black spots. Body is moderately built; head is distinct from neck; eyes are large with a rounded pupil; dorsals are keeled, except smooth outer rows; subcaudals are paired; cloacal scute is divided. **DISTRIBUTION** Myanmar, Thailand, Peninsular Malaysia, Singapore, Sumatra, Borneo, Java. Extralimitally: Laos, Cambodia, Vietnam, E India, Nepal, Bangladesh, China, Sulawesi, Ternate. **HABITS AND HABITAT** Forested lowlands and mid-hills (<1,780m asl). Terrestrial and semi-arboreal, near wetlands. Diet comprises frogs, toads, tadpoles and fish. Oviparous, laying 5–17 eggs (17.5–27 × 11–15mm). **VENOM** Severe envenoming possible, has caused deaths. Anti-Yamakagashi Antivenom (The Japan Snake Institute).

Chinese Keelback Water Snake ■ *Trimerodytes percarinata* 110cm

DESCRIPTION Top of body and forehead are olive-grey, dark brown or black, with 28–36 light-edged black bars on flanks that are broad dorsally and turn narrow laterally; rest of head is yellowish cream; belly is yellowish cream anteriorly, with small black speckling towards posterior; subcaudals are dark grey with black spots. Body is robust; head is large and distinct from neck; eyes are large with a rounded pupil; dorsals are keeled; subcaudals

are paired; cloacal scute is divided. **DISTRIBUTION** Myanmar, Thailand. Extralimitally: Laos, Vietnam, China, E India. **HABITS AND HABITAT** Evergreen hill forests to montane forests, near water bodies (300–2,000m asl). Both aquatic and terrestrial. Diet comprises fish and frogs. Oviparous, laying 4–12 eggs (size unknown).

Yunnan Keelback Water Snake ■ *Trimerodytes yunnanensis* 49.8cm

DESCRIPTION Top of body is brown or brownish black, with transverse black lines that form an X-shaped pattern on flanks; belly is cream. Body is robust; head is slightly distinct from neck; eyes are large with a rounded pupil; dorsals are keeled, except for 1–3 rows of smooth lateral scales; subcaudals are paired; cloacal scute is divided. **DISTRIBUTION** Myanmar. Extralimitally: S China. **HABITS AND HABITAT** Sub-montane to montane forests (900–2,000m asl). Aquatic. Diet and reproductive habits are unstudied.

Yellow-spotted Keelback Water Snake

■ *Xenochrophis flavipunctatus* 97.4cm
(Thai: Ngu Lai-so Suan)

DESCRIPTION Top of body is olivaceous or greenish grey, sometimes with a reddish tinge, and with black spots that develop into a reticulate pattern posteriorly; forehead is dark olive or grey, with 2 dark lines extending from eye, an interrupted band running from base of jaw across neck to form a V-shape, and 2 small median lines; belly is green, cream or yellow, each scale with a black line at posterior edge. Body is robust and cylindrical; head is slightly distinct from neck; eyes are large with a rounded pupil; tail is short; dorsals are weakly keeled anteriorly, distinct posteriorly; subcaudals are paired; cloacal scute is divided. **DISTRIBUTION** Myanmar, Thailand, Peninsular Malaysia. Extralimitally: Laos, Cambodia, Vietnam, E India, China. **HABITS AND HABITAT** Lowlands, including ponds, swamps and flooded rice fields. Diet includes fish and frogs. Oviparous, laying 20–60 eggs (size unknown).

Malayan Spotted Keelback Water Snake

■ *Xenochrophis maculatus* 100cm

DESCRIPTION Top of body is brownish olive, with 4 longitudinal series of small, dark, squarish marks, and a paired row of yellow spots; forehead is dark brown or black; supralabials are yellow with darker smudges; belly is yellow; ventrals are edged with black. Body is slender and cylindrical; head is distinct from neck; eyes are large with a rounded pupil; dorsals are keeled; cloacal scute is divided. **DISTRIBUTION** Peninsular Malaysia, Singapore, Sumatra, Pulau Bangka, Pulau Belitung, Natuna and Riau archipelagos, Borneo. **HABITS AND HABITAT** Open forests in lowlands. Terrestrial and semi-aquatic, near streams and ditches. Diet comprises frogs and toads. Reproductive habits are unstudied.

Javanese Keelback Water Snake ■ *Xenochrophis melanzostus* 97.5cm

DESCRIPTION Top of body may either have elongated blotches on dorsum, or broad, dark, longitudinal stripes on dorsum; has a wide U- or V-shaped mark on nuchal region. Body is robust and cylindrical; head is slightly distinct from neck; eyes are large with a rounded pupil; tail is relatively short in females, compared to related species; dorsals are keeled. **DISTRIBUTION** Java, Bali. **HABITS AND HABITAT** Lowlands, near fresh waters. Diet and reproductive biology unstudied.

Chequered Keelback Water Snake ■ *Xenochrophis piscator* 102cm
(Thai: Ngu Lai-so Yai)

DESCRIPTION Top of body is olive-brown with black spots arranged in 5–6 rows; forehead is brown with a black sub-ocular stripe from eye to supralabial and from post-oculars to angle of jaws; an inverted V-shape mark is present on nuchal region. Body is robust and cylindrical; head is distinct from neck; eyes are large with a round pupil; dorsals are strongly keeled; subcaudals are paired; cloacal scute is divided. **DISTRIBUTION** Myanmar, Thailand. Extralimitally: Afghanistan, Pakistan, India, Sri Lanka, Bangladesh, Nepal, Laos, S China. **HABITS AND HABITAT** Flooded rice fields, ponds, lakes, marshes and rivers. Diet comprises fish and frogs. Oviparous, laying 4–100 eggs (15–40mm).

Red-sided Keelback Water Snake ▪ *Xenochrophis trianguligerus* 135cm
(Bahasa Malaysia: Ular Air. Bahasa Indonesia: Ular Cikopo Merah.
Thai: Ngu Lai-so Lai Sam-laem)

DESCRIPTION Top of body is blackish brown, with orangish-red triangles on sides of neck and front portion of body; bright colours turn olive-brown or grey with age; has dark triangle-shaped marks on top of body; some scales on lips are black-edged; belly is cream. Body is slender and cylindrical; head is large and distinct from neck; eyes are large with a rounded pupil; dorsals are keeled, except for outer 1–2 rows; tail is short; cloacal scute is divided. **DISTRIBUTION** Myanmar, Thailand, Peninsular Malaysia, Singapore, Sumatra, Pulau Nias, Pulau Bangka, Pulau Belitung, Mentawai and Riau archipelagos, Borneo, Java. Extralimitally: Cambodia, Vietnam, Sulawesi, Nicobar Archipelago. **HABITS AND HABITAT** Lowland forests and rice paddies. Semi-aquatic, near streams, standing water bodies, small ditches and puddles (<1,400m asl). Diet comprises frogs. Oviparous, laying 5–15 eggs (15–17 × 29–34mm).

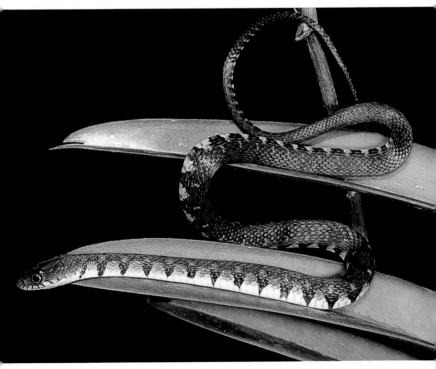

Baram False Cobra ▪ *Pseudoxenodon baramensis* 825mm

DESCRIPTION Top of body greyish-olive, with indistinct dark network; belly yellow, especially towards end; outer margins of subcaudals have pale line. Body robust; head short and distinct from neck; single preocular; nostrils and eyes large; pupils rounded; tail long; dorsal scales keeled. **DISTRIBUTION** Borneo (Sarawak State). **HABITAT AND HABITS** Found in submontane forests at around 1,000m above sea level. Diet and reproductive habits unknown.

Javanese False Cobra ■ *Pseudoxenodon inornatus* 741mm

DESCRIPTION Top of body olive-brown or pale brown. Juveniles have a black nuchal chevron and dark rhomboidal marks with pale centres. Cream-coloured stripe extends along

back of body and tail; belly yellow or mid-brown with dark brown speckles. Body robust; head short and distinct from neck; preoculars number 2–3; nostrils and eyes large; pupils rounded; tail long; dorsal scales keeled. **DISTRIBUTION** Western Java. **HABITAT AND HABITS** Occurs in foothills and on low hills. Diurnal and terrestrial, and found in grassland and nearby tea plantations. Diet and reproductive habits unknown.

Large-eyed False Cobra ■ *Pseudoxenodon macrops* 140cm
(Thai: Ngu Lai-sab Ta To)

DESCRIPTION Top of body is brownish grey, red or olivaceous; has a series of yellow, reddish-brown or orange cross-bars or spots, and a dorso-lateral series of dark spots; nape has a chevron-shaped mark; belly is yellow, with quadrangular black or dark brown spots or cross-bars. Body is stout; head is distinct from neck; eyes are large with a rounded pupil; tail is long; dorsals are keeled, except for the smooth lower rows; cloacal scute is divided.

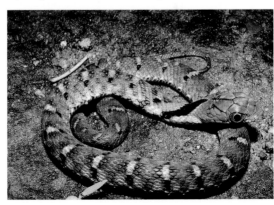

DISTRIBUTION Myanmar, Thailand, Peninsular Malaysia, Borneo. Extralimitally: Laos, Vietnam, E India, Bhutan, Nepal, S China. **HABITS AND HABITAT** Evergreen and dipterocarp forests from mid-hill to montane zones (1,500–2,020m asl). Terrestrial. Diet includes frogs and lizards. Oviparous, producing 6–10 eggs (size unknown).

Elapidae – Cobras and Kraits, Coral and Sea Snakes
The members of this family all produce neurotoxic venom and include many of the most dangerously venomous snakes found on land and in water in Southeast Asia. There are three subfamilies (considered as families by some authors): Elapinae, comprising cobras, kraits and coral snakes, which are mostly permanently terrestrial, have permanently erect fangs, show oviparity and have a distribution in Asia, New Guinea, Australia and Africa; Hydrophiinae, found mostly in marine environments (a few species occur in freshwater habitats), which have a laterally compressed body, oar-like tail, lingual salt glands and greatly reduced ventrals, are ovoviviparous, and are widespread in warmer seas; and Laticaudinae, which are restricted to marine habitats but come ashore to rest, eat and bask, have an oar-like tail, relatively wide ventrals and laterally located nostrils, are oviparous, and have a distribution centred around the Indo-Pacific.

Himalayan Krait ■ *Bungarus bungaroides* 140cm

DESCRIPTION Top of body is black or dark brown, with yellowish-cream transverse lines or narrow bars that form broad bands across venter; has a white stripe across snout, one across nape and another extending from eye to end of jaws. Body is robust; head is indistinct from neck; eyes are rather small with rounded pupils; dorsals are smooth; vertebrals are enlarged; last few subcaudals are paired; cloacal scute is entire. **DISTRIBUTION** Myanmar. Extralimitally: E India, Nepal. **HABITS AND HABITAT** Evergreen and subtropical forests in mid-hills and montane regions (<2,040m asl). Terrestrial. Diet and reproductive habits are unstudied. **VENOM** Neurotoxic; severe envenoming and potentially lethal. SII Polyvalent Antisnake Venom Serum (Serum Institute of India Ltd).

Malayan Krait ■ *Bungarus candidus* 160cm
(Bahasa Malaysia: Ular Katam Tebu. Bahasa Indonesia: Ular Weling. Sundanese: Oraj Weling. Thai: Ngu Tab-sa-ming-khla)

DESCRIPTION Top of body is black or bluish black, usually with 20–35 broad white cross-bars on body that are as wide as or wider than dark interspaces, and 7–10 on tail (populations from Java and Bali, may lack pale cross-bars); indistinct light chevron on nuchal region is sometimes present; supralabials and belly are plain white. Body is robust; head is indistinct from neck; eyes are small with a rounded pupil; tail is short, with an acute tip; dorsals are smooth; subcaudals are single; cloacal scute is entire. **DISTRIBUTION** Thailand, Peninsular Malaysia, Singapore, Sumatra, Pulau Nias, Java, Bawean, Bali. Extralimitally: Laos, Cambodia, Vietnam. **HABITS AND HABITAT** Forested lowlands and sub-montane regions, including near villages and agricultural areas (<1,525m asl). Terrestrial. Diet includes snakes, lizards and toads. Oviparous, clutches comprising 6–10 eggs **VENOM** Neurotoxic; severe envenoming likely. Banded Krait Antivenin (Thai Red Cross Society).

Banded Krait ■ *Bungarus fasciatus* 225cm
(Bahasa Malaysia/Indonesia: Ular Buku Tebu, Ular Katam. Dusun: Tekamas. Iban: Ular Kendawang, Ular Kengkang Tebu, Ular Tetak Tebu. Sundanese: Oraj Welang. Thai: Ngu Sam Lium)

DESCRIPTION Top of body is yellow or pale brown, with black bands approximately equal in size to pale interspaces; forehead has a pale V-shaped marking; belly is pale yellow or brown, with bands. Body is robust, triangular in cross section, with a raised vertebral region; eyes are small; vertebrals are strongly enlarged; tail is blunt-tipped; dorsals are smooth; subcaudals are single; cloacal scute is entire. **DISTRIBUTION** Myanmar, Thailand, Peninsular Malaysia, Singapore, Sumatra, Borneo, Java. Extralimitally: Cambodia, Laos, Vietnam, E India, Nepal, Bangladesh, China. **HABITS AND HABITAT** Light forests, swamps and near villages (<2,500m asl). Terrestrial. Diet comprises snake, lizards, frogs, fish and reptile eggs. Oviparous, clutches comprising 3–14 eggs (22–38mm long). **VENOM** Pre- and post-synaptic neurotoxins; may result in severe envenoming that is potentially lethal. Polyvalent Anti Snake Venom Serum (Central Research Institute), Banded Krait Antivenin (Red Cross Society), Antivenin Polyvalent (Equine) (P.T. Bio Farma Persero, Bandung).

Red-headed Krait ■ *Bungarus flaviceps* 207cm

(Bahasa Malaysia: Ular Katang Kepala Merah. Iban: Ular Kendawang. Thai: Ngu Sam Lium Hua Hang Daeng)

DESCRIPTION Top of body is blue-black with a yellow or tan vertebral stripe; forehead and tail are red, orange or yellow; belly is pink or yellow; in ssp. flaviceps head is red or orange-yellow (sometimes yellow) and body lacks black rings; in ssp. baluensis posterior half of body and tail have 8 pairs of thick black rings, each separated by a narrow white ring. Body is robust and triangular in cross section; head is large and distinct from neck; snout is blunt; eyes are small with a rounded pupil; tail is relatively short; dorsals are smooth;

cloacal scute is entire. **DISTRIBUTION** Myanmar, Thailand, Peninsular Malaysia, Sumatra, Pulau Belitung, Borneo, Java. Extralimitally: Cambodia, Vietnam. **HABITS AND HABITAT** Ssp. *flaviceps*, forested lowlands and mid-hills (<914m asl); ssp. *baluensis*, sub-montane forests (550–900m asl). Terrestrial. Diet comprises snakes and lizards. Reproductive biology is unstudied. **VENOM** Pre- and post-synaptic neurotoxins; may result in severe envenoming that is potentially lethal. Banded Krait Antivenin (Thai Red Cross Society).

Many-banded Krait ■ *Bungarus multicinctus* 135.4cm

DESCRIPTION Top of body is jet-black or bluish black, with 27–44 light cross-bars on body and 7–17 on tail; pale bands expand on flanks; supralabials and belly are yellowish cream; subcaudals are mottled with dark brown. Body is robust; head is indistinct from neck; eyes are small with a rounded pupil; dorsals are smooth; tail is long, thin and tapering; subcaudals are entire; cloacal scute is entire. **DISTRIBUTION** Myanmar.

Extralimitally: Laos, Vietnam, E China. **HABITS AND HABITAT** Lowlands and sub-montane forests, especially near wetlands, including agricultural areas. Terrestrial. Diet includes snakes, rodents, frogs, lizards and eels. Oviparous, laying 3–12 eggs (size unknown). **VENOM** Pre- and post-synaptic neurotoxins; may result in severe envenoming that is potentially lethal. Purified *Bungarus multicinctus* Antivenom (Shanghai Institute of Biological Products); Bivalent Antivenom Elapid, Naja-Bungarus Antivenin (National Institute of Preventative Medicine, Taiwan).

Black Krait ■ *Bungarus niger* 120cm

DESCRIPTION Top of body is black, brown or bluish black; infralabials are cream; belly and undersurface of tail are white with distinct dark mottling. Body is robust; head is indistinct from neck; eyes are small with a rounded pupil; tail tapers to a point; mid-dorsal row of vertebrals is strongly enlarged, as broad as or broader than long; dorsals are smooth; subcaudals are undivided; cloacal scute is entire. **DISTRIBUTION**. Myanmar. Extralimitally: E India, Bhutan, Nepal. **HABITS AND HABITAT** Evergreen forests and forest edges, including agricultural fields (300–1,450m asl). Terrestrial. Diet includes small snakes. Reproductive habits are unstudied. **VENOM** Pre- and post-synaptic neurotoxins; may result in severe envenoming that is potentially lethal. No antivenom raised.

Blue Coral Snake ■ *Calliophis bivirgatus* 185cm
(Bahasa Malaysia: Ular Pantai Biru Biru, Ular Sina Mata Hari. Bahasa Indonesia: Ular Tjabeh. Iban: Ular Siran, Ular Kendawan. Thai: Ngu Prik Tong Daeng)

DESCRIPTION Top of body is dark blue or blue-black, with distinct stripes along flanks; head, tail and belly are coral-red. The subspecies may be identified on the basis of additional colours and patterns: *bivirgatus* has a narrow white paravertebral stripe; *flaviceps* has a pale blue paravertebral stripe; *tetrataenia* has 4 lateral stripes, the outermost broadest. Body is slender; head is short and indistinct from neck; eyes are small with a rounded pupil; tail is short, terminating in a sharp point; dorsals are smooth; cloacal scute is divided.

DISTRIBUTION Myanmar, Thailand, Peninsular Malaysia, Singapore, Sumatra, Pulau Nias, Pulau Bangka, Mentawai and Riau archipelagos, Borneo, Java. Extralimitally: Laos, Cambodia. **HABITS AND HABITAT** Lowland to sub-montane forests (<1,200m asl); also forest fringes, such as in agricultural areas. Diet includes other snakes. Oviparous, clutches comprising 1–3 eggs (35–35 × 9mm). **VENOM** Unknown, and potentially lethal envenoming. No antivenom raised.

Malayan Striped Coral Snake ■ *Calliophis intestinalis* 71cm
(Bahasa Malaysia/Indonesia: Ular Cabe, Ular Tali Kasut. Iban: Ular Siram. Thai: Ngu Prik Si Namtan)

DESCRIPTION In ssp. *intestinalis*, dorsum is black; vertebral region and lower flanks have narrow white lines; mid-flanks have a dark brown stripe; belly has broad black and white transverse bands, the black cross-bands in contact with black on sides; forehead is black with a Y-shaped cream mark. In ssp. *lineata*, dorsum is greyish brown; vertebral region and flanks have narrow, black-edged white lines, these extending to tail tip; lower flanks have a separate narrow, black-edged white line; belly has broad black and white transverse bands, the black cross-bands not in contact with black on sides; forehead is brown, mottled with black; subcaudals are orange, with 2 narrow bars. In ssp. *thepassi*, dorsum is black; broad black vertebral stripe extends to tail; dorso-lateral stripes are 2 scales wide; belly has broad black and white transverse bands, the black cross-bands not in contact with black on sides; forehead is rufous brown. Body is slender; head is small and indistinct from neck; dorsals are smooth; cloacal scute is entire. **DISTRIBUTION** Thailand, Peninsular Malaysia, Singapore, Sumatra, Pulau Nias, Pulau Bangka, Pulau Belitung, Mentawai and Riau archipelagos, Borneo, Sumatra, Java. Extralimitally: Cambodia, Vietnam, Palawan, Balabac, Busuanga, Luzon, Mindanao, Samar, Sulu Archipelago. **HABITS AND HABITAT** Lowland to sub-montane forests, plus parks and gardens (<1,100m asl). Terrestrial and semi-fossorial, concealing itself under debris on the forest floor by day. Diet comprises other snakes. Oviparous, producing 1–3 eggs (27–35 × 8–9mm). **VENOM** Unknown, and potentially lethal envenoming. No antivenom raised.

ABOVE: *dorsum*. BELOW: *belly*

Speckled Coral Snake ■ *Calliophis maculiceps* 48cm
(Bahasa Malaysia: Ular Pantai Bintik Kecil. Thai: Ngu Plong-wai Hua Dam)

DESCRIPTION Top of body is brownish yellow or reddish brown, with or without black spots in a longitudinal series along each side; head and nape are black with a yellow occipital spot; pale band across forehead may be present or absent; supralabials are yellow; belly is pale blue or grey. Body is slender; head is small and indistinct from neck; eyes are small with a rounded pupil; tail is short; dorsals are smooth; mid-dorsals are not enlarged; subcaudals are paired; cloacal scute is divided. **DISTRIBUTION** Myanmar, Thailand, Peninsular Malaysia. Extralimitally: Laos, Cambodia, Vietnam. **HABITS AND HABITAT** Lowland and sub-montane forests and plantations (<1,330m asl). Terrestrial, on edges of streams. Diet comprises small snakes. Oviparous, producing clutches of 2 eggs (size unknown). **VENOM** Unknown, and potentially lethal envenoming. Polyvalent Anti Snake Venom Serum (Central Research Institute, Kasauli).

Annandale's Sea Snake ■ *Hydrophis annandalei* 52cm
(Thai: Ngu Chai-thong Hua To)

DESCRIPTION Top of body is grey-blue with 35–46 dark bands, these broader than their pale interspaces; head is olivaceous; belly is pale yellow or cream. Body is short and robust; head is small and narrower than widest part of body; forehead scales are fragmented; eyes are small with a rounded pupil; tail is flattened; dorsals are keeled; cloacal scutes are enlarged, bordered with 1–2 small scales. **DISTRIBUTION** Thailand, Peninsular Malaysia, Singapore, Borneo, possibly Sumatra. Extralimitally: Cambodia, Vietnam. **HABITS AND HABITAT** Sea coasts, probably in shallow waters with a sandy bottom; also fresh waters. Diet comprises fish. Reproductive habits are unstudied. **VENOM** Postsynaptic neurotoxins, clinical effects uncertain and may result in major envenoming. Sea Snake Antivenom (CSL Limited, Victoria).

Blue-grey Sea Snake ■ *Hydrophis caerulescens* 109cm
(Thai: Ngu Samae-rang Lai Yoeng)

DESCRIPTION Top of body is bluish white or bluish grey, with 40–60 dark bands that narrow on lower flanks; forehead is dark (juveniles with a yellow or cream U-shaped mark); belly is cream, with bands twice as wide as interspaces, tapering ventrally. Body is moderately robust; head is small, with projecting upper jaw; eyes are small with a rounded

pupil; scales are quadrangular or hexagonal, weakly imbricate or juxtaposed; tail is flattened; dorsals are keeled; ventrals are less than twice as large as adjacent scales; cloacal scute is divided. **DISTRIBUTION** Myanmar, Thailand, Peninsular Malaysia, Singapore, Sumatra, Borneo. Range extends from Pakistan to N Australia. **HABITS AND HABITAT** Shallow seas, estuaries and an inland lake (Songkhla Lake, S Thailand). Diet comprises burrowing gobies. Ovoviviparous, producing 2–6 neonates (size unknown). **VENOM** Postsynaptic neurotoxins, clinical effects uncertain and may result in major envenoming. Sea Snake Antivenom (CSL Limited, Victoria).

Short Sea Snake ■ *Hydrophis curtus* 97.2cm
(Bahasa Indonesia: Ular Lempe. Thai: Ngu Ai-ngua)

DESCRIPTION In adults, top of body is typically plain olive to dark grey; in juveniles, top of body is brownish grey to olive, with 35–55 olive to dark grey bands that taper to a point on flanks; belly is plain yellow or cream, or has a narrow, dark ventral stripe or irregular band. Body is robust and short; head is broad and short; scales are squarish or hexagonal,

lowermost rows (especially in males) with a short keel; head shields are entire; nostril is dorsally situated; eyes are small with a rounded pupil; tail is flattened; cloacal scutes are weakly enlarged. **DISTRIBUTION** Myanmar, Thailand, Peninsular Malaysia, Singapore, Sumatra, Borneo. Extralimitally: from Persian Gulf to Australia. **HABITS AND HABITAT** Coasts with muddy bottoms; also coral reefs (depths of 6–15m). Diet comprises eels, gobies, catfish, squid and other marine invertebrates. Ovoviviparous, producing 1–6 neonates (250–373mm). **VENOM** Postsynaptic neurotoxins; possibility of severe and lethal envenoming. Sea Snake Antivenom (CSL Limited, Victoria).

Annulated Sea Snake ■ *Hydrophis cyanocinctus* 188.5cm
(Bahasa Indonesia: Ular Laut Besar Berdagu Kehitaman. Thai: Ngu Samae-rang Tong Luang Lai-kram)

DESCRIPTION Top of body is typically olive or yellow, with numerous bluish-black transverse bands that may encircle body; head is yellowish green; belly is yellowish

cream. Body is elongate and moderately robust anteriorly, thickening posteriorly; head is small, almost indistinct from neck; eyes are small with a rounded pupil; dorsals are strongly keeled; tail is flattened; cloacal scutes are enlarged. **DISTRIBUTION** Thailand, Peninsular Malaysia, Singapore, Borneo. Extralimitally: Persian Gulf to E China, Vietnam, Philippines and Japan. **HABITS AND HABITAT** Shallow coastal waters. Diet comprises gobies, eels and marine invertebrates. Ovoviviparous, producing 3–16 neonates (381mm). **VENOM** Postsynaptic neurotoxins; potential for severe and lethal envenoming. Sea Snake Antivenom (CSL Limited, Victoria).

Banded Sea Snake ■ *Hydrophis fasciatus* 110cm
(Bahasa Indonesia: Ular Lait. Bahasa Malaysia: Ular Selimpat. Iban: Ular Birang. Thai: Ngu Samae-hua Dam)

DESCRIPTION Top of body and forehead are dark olive to black, with pale yellow oval spots on flanks that may be connected as cross-bars; posterior is grey; belly is cream; dark rhomboidal spots along flanks may form annuli in juveniles. Body is slender anteriorly,

thickening posteriorly; head is small; eyes are small with a rounded pupil; scales are juxtaposed or slightly imbricate; tail is flattened; cloacal scutes are divided. **DISTRIBUTION** Myanmar, Thailand, Peninsular Malaysia, Singapore, Sumatra, Borneo, Java. Extralimitally: India, Pakistan, Vietnam. **HABITS AND HABITAT** Shallow coastal waters. Diet comprises anguiliform eels. Reproductive habits are unstudied. **VENOM** Postsynaptic neurotoxins; potential for severe and lethal envenoming. Sea Snake Antivenom (CSL Limited, Victoria).

Narrow-headed Sea Snake ▪ *Hydrophis gracilis* 95cm
(Thai: Ngu Ko-on Hua Khem)

DESCRIPTION Top of body is yellow with 40–60 black bands or lateral blotches; forehead is black; supralabials are cream; neck has dark bands with narrow, light grey interspaces; belly is cream. Body is slender anteriorly, thickening posteriorly; head is small, with projecting upper jaws; eyes are small with a rounded pupil; neck is slender; juxtaposed

scales are present on thickest part of body; tail is flattened; dorsals are keeled; ventrals are divided by a longitudinal fissure; cloacal scutes are weakly enlarged. **DISTRIBUTION** Myanmar, Thailand, Vietnam, Peninsular Malaysia, Singapore, Sumatra, Borneo, Java. Extralimitally: from Persian Gulf to Australia and Melanesia. **HABITS AND HABITAT** Deep, turbid coastal waters; possibly a bottom-dweller. Diet includes anguiliform eels. Ovoviviparous, producing up to 6 neonates (size unknown). **VENOM** Postsynaptic neurotoxins; potential for severe and lethal envenoming. Sea Snake Antivenom (CSL Limited, Victoria).

Obscure-patterned Sea Snake ▪ *Hydrophis obscurus* 119cm

DESCRIPTION In adults, top of body is plain grey; in juveniles, top of body is black or bluish black, with 30–60 yellow or cream cross-bars that encircle body and tail; forehead

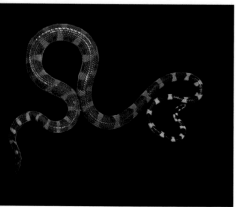

has a curved yellow marking between snout and sides of parietals; belly is yellow. Body is robust and elongate; scales are rounded with blunt or obtusely pointed tips, and are imbricate; eyes are small with a rounded pupil; tail is flattened; cloacal scutes are moderately enlarged. **DISTRIBUTION** Myanmar. Extralimitally: India. **HABITS AND HABITAT** Brackish-water sections of rivers, large coastal lagoons and salt-water lakes. Diet and reproductive habits are unstudied. **VENOM** Postsynaptic neurotoxins; potential for severe and lethal envenoming. Sea Snake Antivenom (CSL Limited, Victoria).

Ornate Sea Snake ■ *Hydrophis ornatus* 115cm
(Bahasa Indonesia: Ular Laut Berlurik. Thai: Ngu Samae-rang Hang Khao)

DESCRIPTION Top of body is light or greyish olive to plain cream, with 30–60 broad, dark bars or rhomboidal spots separated by narrow interspaces on body, and 6–11 on tail; belly is yellow or cream. Body is robust; head is large; eyes are small with a rounded pupil; scales are hexagonal, and feebly imbricate or juxtaposed; tail is flattened; dorsals are tuberculate or have a short keel; anteriorly, ventrals are twice as large as adjacent scales; cloacal scutes are weakly enlarged.

DISTRIBUTION Myanmar, Thailand, Peninsular Malaysia, Singapore, Sumatra, Borneo. Extralimitally: from Persian Gulf, E to Vietnam, Philippines, New Guinea and Australia. **HABITS AND HABITAT** Shallow waters with coral reefs, river mouths and estuaries. Diet includes fish, especially marine catfish. Ovoviviparous, producing 1–4 neonates (c. 34mm). **VENOM** Postsynaptic neurotoxins; potential for severe and lethal envenoming. Sea Snake Antivenom (CSL Limited, Victoria).

Horned Sea Snake ■ *Hydrophis peronii* 125cm
(Bahasa Indonesia: Ular Laut Bertanduk. Thai: Ngu Tak-lai Tong Kao)

DESCRIPTION Top of body is light brown, with dark bands encircling body, widest on vertebral region and narrowing ventrally; forehead is pale brown; belly is paler. Body is

robust; head is small with symmetrical shields, some with spines on their posterior edges; eyes are small with a rounded pupil; tail is flattened; dorsals are keeled; cloacal scute is entire. **DISTRIBUTION** Thailand, Peninsular Malaysia, Singapore. Extralimitally: Vietnam, S China, N Australia, New Caledonia. **HABITS AND HABITAT** Shallow seas, including reefs at medium depth, as well as estuaries. Diet comprises burrowing gobies. Ovoviviparous, producing 10 neonates (size unknown). **VENOM** Postsynaptic neurotoxins; potential for severe and lethal envenoming. Sea Snake Antivenom (CSL Limited, Victoria).

Yellow-bellied Sea Snake ■ *Hydrophis platura* 100cm
(Thai: Ngu Chai-thong Lang Dam)

DESCRIPTION Top of body is black or dark brown; belly is light brown or yellow, sometimes with a series of black spots or bars; dorsal and ventral colours are sharply separated; tail has a bright yellow diamond-shaped pattern. Body is slender and compressed; head is elongate, bill-like, slightly flattened and distinct from neck; eyes are small with a rounded pupil; tail is flattened; ventrals are irregular in shape and indistinct beyond anterior of body; cloacal scutes are enlarged. **DISTRIBUTION** Myanmar,

Thailand, Peninsular Malaysia, Singapore, Sumatra, Borneo, Java. Extralimitally: Pacific and Atlantic oceans. **HABITS AND HABITAT** Pelagic species, known mostly from sea coast strandings. Diet comprises surface-feeding fish. Ovoviviparous, producing clutches of 2–6 neonates (220–260mm). **VENOM** Postsynaptic neurotoxins; possibility of severe and lethal envenoming. Sea Snake Antivenom (CSL Limited, Victoria).

Beaked Sea Snake ■ *Hydrophis schistosa* 158cm
(Bahasa Malaysia: Ular Selimpat. Bahasa Indonesia: Ular Laut Berparuh. Thai: Ngu Ko-on Pak Cha-ngoi)

DESCRIPTION In adults, top of body is greyish olive or silvery grey, with a darker forehead; body has indistinct darker markings, sometimes forming 40–60 dark transverse bands; suborbital stripe is absent; belly is cream anteriorly, darkening to greenish yellow towards tail. Juveniles are dark grey on top. Body is robust; head is narrow, nearly indistinct from neck; eyes are small with a rounded pupil; rostral extends ventrally; dorsals are keeled; tail is

flattened; cloacal scutes are slightly enlarged. **DISTRIBUTION** Myanmar, Thailand, Peninsular Malaysia, Singapore, Sumatra, Borneo. Range extends from Persian Gulf, through S Asia, E to New Guinea and N Australia. **HABITS AND HABITAT** Shallow (less than 5m) sea coasts and mangroves; may travel upriver into freshwater habitats. Diet comprises marine catfish. Ovoviviparous, producing 4–33 neonates (150–280mm). **VENOM** Postsynaptic neurotoxins; possibility of severe and lethal envenoming. Sea Snake Antivenom (CSL Limited, Victoria).

Sibau River Sea Snake ▪ *Hydrophis sibauensis* 73.5cm

DESCRIPTION Top of body is grey-brown, darkening posteriorly, with 49–58 incomplete yellow to light orange bands; forehead is black with small yellow spots and arrow-shaped markings; belly, including throat, is black anteriorly and greyish yellow from mid-body to posterior. Body is slender; head is narrower than body; nostril is on top of head; forehead shields are large and regular; eyes are small with a rounded pupil; tail is flattened; dorsals have a median keel; ventrals are small and distinct throughout; cloacal scutes are enlarged. **DISTRIBUTION** Borneo. **HABITS AND HABITAT** Known only from a freshwater section of Sungai Sibau, a tributary of Sungai Kapuas more than 1,000km upriver. Diet is unstudied. Ovoviviparous, producing 7 neonates (size unknown). **VENOM** Presumably with postsynaptic neurotoxins; possibility of severe and lethal envenoming. Sea Snake Antivenom (CSL Limited, Victoria).

Spiral Sea Snake ▪ *Hydrophis spiralis* 275cm
(Thai: Ngu Samae Lai-luang)

DESCRIPTION In adults, top of body and forehead are olive-yellow to olive-brown, with 35–50 encircling dark bands that are narrower than interspaces, sometimes with a black dorsal spot between each; in juveniles, forehead is nearly black with a yellow horseshoe-shaped mark; flanks are yellowish cream; belly is yellow or cream; tip of tail is dark brown or black. Body is elongate and moderately robust anteriorly, thickening posteriorly; head and neck are slender; eyes are small with a rounded pupil; tail is flattened; dorsals are smooth or keeled; ventrals are distinct throughout, about twice as broad as adjacent body scales; cloacal scutes are enlarged. **DISTRIBUTION** Myanmar, Thailand, Peninsular Malaysia, Singapore, Sumatra, Borneo. Extralimitally: from Persian Gulf to Sulawesi and Philippines. **HABITS AND HABITAT** Deep-water habitats (more than 10m). Diet is unstudied. Ovoviviparous, producing 5–14 neonates (200–405mm). **VENOM** Postsynaptic neurotoxins; possibility of severe and lethal envenoming. Sea Snake Antivenom (CSL Limited, Victoria).

Grey Sea Snake ■ *Hydrophis viperina* 92.5cm
(Bahasa Indonesia: Ular Laut Abu-abu. Thai: Ngu Chai-thong Tong Kao)

DESCRIPTION Top of body is uniform grey, or with lighter mottling or 25–35 dark cross-bars or spots; forehead is grey or black, sometimes with dark mottling; belly is cream or white; juveniles have a large, dark vertebral pattern. Body is robust; head is short and depressed; forehead scales are entire; nostrils are on dorsal surface; eyes are small with a rounded pupil; scales on body are hexagonal and keeled; tail is flattened; ventrals at anterior cover half body width, narrowing posteriorly to twice width of adjacent scales or less; has 4 cloacal scutes, the outer 2 largest. **DISTRIBUTION** Myanmar, Thailand, Peninsular Malaysia, Singapore, Borneo. Extralimitally: from Persian Gulf to E China. **HABITS AND HABITAT** Seas up to 32km from the coast; also river mouths. Diet consists of marine invertebrates and eels. Ovoviviparous, producing 3–5 neonates **VENOM** Postsynaptic neurotoxins; possibility of severe and lethal envenoming. Sea Snake Antivenom (CSL Limited, Victoria).

Yellow-lipped Sea Krait ■ *Laticauda colubrina* 171cm
(Bahasa Indonesia: Krait Laut Berbibir Kuning. Thai: Ngu Saming-talay Pak-luang)

DESCRIPTION Top of body is blue-grey with 24–64 black annuli; supralabials are yellow; belly is cream. Body is robust (especially in adult females) and cylindrical; nostrils are lateral; rostral is undivided; nasals are separated by internasals; eyes are small with a rounded pupil; tail is flattened; dorsals are smooth; cloacal scute is divided. **DISTRIBUTION** Myanmar, Thailand, Peninsular Malaysia, Singapore, Sumatra, Pulau Belitung, Borneo, Java, Bali. Extralimitally:

widespread in the Indian and Pacific oceans, E to Polynesia. **HABITS AND HABITAT** Shallow seas, around rocky islets, to depths of *c.* 60m. Comes ashore to lay eggs, bask, rest and digest food. Diet includes anguiliform eels. Oviparous, producing clutches of 3–13 eggs (44.6–92.2 × 20.3–31.1mm). **VENOM** Postsynaptic neurotoxins; possibility of severe and lethal envenoming. Sea Snake Antivenom (CSL Limited, Victoria).

Monocled Cobra ■ *Naja kaouthia* 230cm
(Bahasa Malaysia: Ular Senduk. Thai: Ngu Hao Mo)

DESCRIPTION Top of body is brown, greyish brown, blackish brown or pale yellow, some individuals with darker bands; hood marking is typically a light circle or mask-shaped with a dark centre (1–2 dark spots are sometimes present in pale oval portion); has a light throat colour with paired lateral spots; rest of underparts are similar to dorsum or have dark pigmentation towards tail; subcaudals are dark-edged. Body is

robust; head is large, distinct from neck; a cuneate is usually present; hood is rounded; eyes are moderate with a rounded pupil; tail is short; dorsals are smooth and glossy; subcaudals are paired; cloacal scute is entire. **DISTRIBUTION** Myanmar, Thailand, Peninsular Malaysia. Extralimitally: Laos, Cambodia, Vietnam, India, Nepal, Bangladesh, S China. **HABITS AND HABITAT** Lowland and mid-hill forests, plus agricultural fields and plantations (<820m asl). Largely terrestrial, but known to swim in lakes and rivers. Diet comprises rodents, frogs, fish and snakes. Oviparous, clutches comprising 15–30 eggs (50–55mm long). **VENOM** Postsynaptic neurotoxins; possibility of severe and lethal envenoming. Cobra Antivenin (Thai Red Cross Society); *Naja kaouthia* Antivenom, University of Medicine and Pharmacy, Ho Chi Minh City); Bivalent (Pharmaceutical Industries Corporation, Yangon); Anti-Cobra, Siamese Cobra (Pharmaceutical Industries Corporation, Yangon).

Indo-Chinese Spitting Cobra ■ *Naja siamensis* 160cm
(Thai: Ngu Hao Ki-raun)

DESCRIPTION Top of body is variable, having either a contrasting black and white pattern, or being black or grey with white speckling, dark with light cross-bars or, occasionally, plain light or dark brown; hood marking is absent, or can be U-, V- or H- shaped; belly is white, with or without black or brown bars. Body is robust; head

is large and distinct from neck; snout is rounded; a single cuneate is present; hood is oval; eyes are moderate with a rounded pupil; tail is short; dorsals are smooth; cloacal scute is entire. **DISTRIBUTION** Thailand. Extralimitally: Laos, Cambodia, Vietnam. **HABITS AND HABITAT** Dry plains and low hills. Diet comprises small mammals. Reproductive biology is unstudied. **VENOM** Postsynaptic neurotoxins; possibility of severe and lethal envenoming. No antivenom raised.

Equatorial Spitting Cobra ■ *Naja sputatrix* 150cm
(Bahasa Indonesia: Ular Hitam, Lipi Oh, Ular Sendok. Sundanese: Oraj Bedul)

DESCRIPTION Top of body is blackish grey, silvery or brown; hood pattern is chevron-, mask-, horseshoe- or spectacle-shaped, or plain; belly is yellowish cream, sometimes with faint scattered spots. Body is robust; head is large and distinct from neck; snout is rounded; hood is elongate; eyes are moderate in size with a rounded pupil; tail is short; dorsals are smooth; subcaudals are divided; cloacal scute is entire. **DISTRIBUTION** Java,

Bali. Extralimitally: Lombok, Alor, Sumbawa, Komodo, Flores, Lembata. **HABITS AND HABITAT** Dry, rocky, lowland deciduous forests (altitude unknown). Both terrestrial and arboreal. Diet comprises toads, rodents and reptiles. Oviparous, clutches comprising 16–26 eggs (40–51 × 23–31mm). **VENOM** Postsynaptic neurotoxins; possibility of severe and lethal envenoming. Cobra Antivenin (Thai Red Cross Society); Antivenin Polyvalent (Equine) (P.T. Bio Farma Persero, Bandung).

Sumatran Cobra ■ *Naja sumatrana* 150cm

(Bahasa Malaysia/Indonesia: Ular Senduk, Ular Senduk Sembur. Iban: Ular Belalang. Thai: Ngu Hao Tong Pon Pit)

DESCRIPTION Coloration of top of body is variable, depending on locality and size; belly of all populations is dark brown or brownish grey. Adults of Peninsular Malaya and Bornean population are bluish black with a plain hood; juveniles have pale, narrow cross-bars on dorsum, and a pale throat with lateral spots and, often, a median spot. Adults of Sumatran population are pale to dark brown with a plain hood; juveniles have *c.* 12 pale, narrow cross-bars on dorsum. Bornean snakes are metallic bluish black with a plain hood; *c.* 12 narrow, chevron-shaped white or yellowish-cream cross-bars; pale throat; bright yellow anterior ventrals. Body is robust; head is large and distinct from neck; snout is rounded; hood is rounded in adults, more elongate in juveniles; eyes are moderate in size with a rounded pupil;

tail is short; dorsals are smooth; cloacal scute is entire. **DISTRIBUTION** Peninsular Thailand, Peninsular Malaysia, Singapore, Sumatra, Pulau Bangka, Pulau Belitung, Riau Archipelago, Borneo. Extralimitally: Palawan, Calamianes Archipelago. **HABITS AND HABITAT** Lightly forested areas in lowlands and mid-hills. Diet includes rats and other small vertebrates. Reproductive habits are unstudied. **VENOM** Postsynaptic neurotoxins; possibility of severe and lethal envenoming. No antivenom raised.

King Cobra ▪ *Ophiophagus hannah* 585cm

(Bahasa Malaysia: Ular Tedung Selar. Bahasa Indonesia: King Kobra. Iban: Ular Belalang. Sundanese: Oraj Totok. Thai: Ngu Chong-ang)

DESCRIPTION In adults, top of body is brownish black, nearly unpatterned to distinctly banded with 27–84 yellow bands, depending on population; scales on posterior and on tail are lighter in middle; chin and throat are yellow, belly is dark grey. In juveniles, top of body is dark brown or black with pale yellow or orange bands; belly is bright anteriorly, yellow or orangish yellow, turning grey beyond throat region. Body is robust in adults, more slender in juveniles; head is large in adults, distinct from neck; paired occipital scales are present; eyes are large; dorsals are smooth; subcaudals are paired or single, or a combination; cloacal scute is entire. **DISTRIBUTION** Myanmar, Thailand, Peninsular Malaysia, Singapore, Sumatra, Pulau Simeulue, Pulau Galang, Pulau Bangka, Pulau Belitung, Borneo, Java, Bali. Extralimitally: E Pakistan, India, Bhutan, Nepal, China, Philippines, Cambodia, Laos,

Vietnam. **HABITS AND HABITAT** Semi-evergreen, evergreen, moist deciduous and tropical dipterocarp forests and mangrove swamps (<2,181m asl). Terrestrial; juveniles are more arboreal. Diet includes other snakes and monitor lizards. Oviparous, laying clutches of 20–43 eggs (50–65mm long) on mound nests of fallen leaves. **VENOM** Postsynaptic neurotoxins; possibility of severe and lethal envenoming. Polyvalent Anti Snake Venom Serum (Central Research Institute, Kasauli); King Cobra Antivenin (Thai Red Cross Society); *Ophiophagus hannah* Antivenom (Venom Research Unit, University of Medicine and Pharmacy, Ho Chi Minh City); SII Polyvalent Antisnake Venom Serum (lyophilized) (Serum Institute of India Ltd.).

MacClelland's Coral Snake ■ *Sinomicrurus macclellandi* 84cm
(Thai: Ngu Plong-wai Lai Kwan Dam)

DESCRIPTION Top of body is reddish brown with 23–40 narrow yellow or pale brown-edged black stripes; forehead is black with a cream or yellow band behind eyes; tail has 2–6 black bands; black transverse bars are reduced to transverse vertebral spots in some individuals; belly is yellowish cream with black bands or squarish marks. Body is slender and cylindrical; head is short and rounded; vertebrals are not enlarged; eyes are small with a rounded pupil; dorsals are smooth; tail is short, with a pointed tip; cloacal scute is divided. **DISTRIBUTION** Myanmar, Thailand. Extralimitally: Laos, Vietnam, E India, Nepal, Bangladesh, China. **HABITS AND HABITAT** Temperate evergreen forests (55–2,500m asl). Terrestrial and sub-fossorial, sheltering under loose soil or in vegetation. Diet comprises other snakes and lizards. Oviparous, clutches comprising 6–14 eggs (20–33.3 × 10.9–12mm). **VENOM** Unknown with possibility of severe and lethal envenoming. Polyvalent Anti Snake Venom Serum (Central Research Institute, Kasauli).

HOMALOPSIDAE – PUFF-FACED WATER SNAKES
This family includes water snakes of the Asia–Pacific region that have reduced eyes and valvular nostrils, features reflecting a life in muddy and/or turbid habitats such as fresh and brackish waters, including flooded agricultural fields, marshes and ponds. They feed on fish, frogs and crustaceans, and are ovoviviparous, producing live young underwater. Homalopsidae species range from the Indian sub-continent in the west, to Southeast Asia, northern Australia and Micronesia.

Yellow-banded Mangrove Snake ▪ *Cantoria violacea* 120cm
(Thai: Ngu Pak Kwang Lai)

DESCRIPTION Top of body is dark blackish grey or black, with dull yellow transverse bars 2–3 scales wide and narrower than their interspaces; head is white-spotted or with 2 yellow cross-bars; belly is plain cream or with grey markings. Body is slender, elongate and cylindrical; head is indistinct from neck; pre-frontals are in broad contact; eyes are small with a rounded pupil; tail is slightly compressed, short and blunt; dorsals are smooth; subcaudals are paired; cloacal scute is divided. **DISTRIBUTION** Myanmar, Thailand, Peninsular Malaysia, Singapore, Sumatra, Borneo. Extralimitally: Andaman Islands, Timor. **HABITS AND HABITAT** Tidal rivers, including estuaries; also mudflats at low tide and mangrove swamps. Aquatic. Diet comprises snapping shrimps and fish. Ovoviviparous (numbers and size of neonates unknown).

Schneider's Dog-faced Water Snake ■ *Cerberus schneiderii* 127cm
(Bahasa Malaysia: Ular Air Kadut, Ular Birang. Bahasa Indonesia: Ular Tambak. Thai: Ngu Pak Kwang Nam Khem)

DESCRIPTION Top of body is dark grey or greyish green with faint, dark blotches; a dark post-ocular stripe reaches sides of neck; belly is yellowish cream with dark grey areas. Body is moderately robust; head is long and distinct from neck; eyes are small and beady with a rounded pupil; tail is short and slender; dorsals are strongly keeled; subcaudals are paired; cloacal scute is divided. **DISTRIBUTION** Myanmar, Thailand, Peninsular Malaysia, Singapore, Sumatra, Mentawai and Natuna archipelagos, Borneo, Java, Bali. Extralimitally: Pakistan, India, Bangladesh, Sri Lanka, Vietnam, New Guinea, Australia. **HABITS AND HABITAT** Low-lying coastal areas, such as mangrove mudflats and rice fields. Diet comprises mudskippers, gobies, crabs and frogs. Ovoviviparous, producing 5–38 neonates (size unknown).

Rainbow Water Snake ■ *Enhydris enhydris* 88.2cm
(Bahasa Malaysia/Indonesia: Ular Air Biasa. Thai: Ngu Sai-rung, Ngu Pla)

DESCRIPTION Top of body is greyish brown or olive-green, with a dark vertebral and 2 light lateral stripes running from upper surface of head to tail; belly is yellowish cream, ventrals with a dark spot on edges creating a dark line on flanks. Body is robust and sub-cylindrical; head is small, depressed and slightly distinct from neck; snout is rounded; nostrils are situated on upper surface of head; eyes are small with a vertical pupil; dorsals are smooth; tail is short and tapering; subcaudals are paired; cloacal scute is divided. **DISTRIBUTION** Myanmar, Thailand, Peninsular Malaysia, Singapore, Sumatra, Borneo, Java. Extralimitally: E India, Nepal, Bangladesh, Bhutan, Vietnam, S China. **HABITS AND HABITAT** Fresh- and brackish-water habitats, including wet rice fields. Diet comprises fish, frogs, tadpoles and lizards. Ovoviviparous, producing 4–20 neonates (158–206mm).

Jagor's Water Snake ▪ *Enhydris jagorii* 56cm
(Thai: Ngu Sai-rung Lai Kwang)

DESCRIPTION Top of body is greyish brown, with black ocellate spots arranged in a linear series between neck and tail; occasionally, dorsum is cream or pink in region below spots; supralabials are yellowish cream; gular region is lavender-brown; belly is pale, its edges marked with a grey zigzag line and indistinct series of dots medially. Body is robust and sub-cylindrical; head is short, rounded and indistinct from neck; eyes are small with a vertical pupil; tail is short and compressed; dorsals are smooth; subcaudals are paired; cloacal scute is divided. **DISTRIBUTION** Thailand. **HABITS AND HABITAT** Freshwater swamps in the Chao Phraya drainage basin. Aquatic. Diet and reproductive habits are unstudied.

Tentacled Snake ▪ *Erpeton tentaculatus* 77cm
(Thai: Ngu Kra Daeng)

DESCRIPTION Top of body is olive, grey or brown, with 2 indistinct, dark longitudinal paravertebral stripes, or variegated black; vertebral region has dark spots or cross-bars; a broad, dark lateral stripe extends from snout, across orbit and along flanks; belly is yellowish brown. Body is slender; head is small and distinct from neck; has a long, scaly,

flexible rostral appendage, as long as snout length; eyes are small and protruding, with a vertical pupil; dorsals are strongly keeled; ventrals are narrow; subcaudals are paired; cloacal scute is entire. **DISTRIBUTION** Thailand. Extralimitally: Cambodia, Vietnam. **HABITS AND HABITAT** Lowland ponds and slow-moving bodies of water. Aquatic. Diet comprises fish. Ovoviviparous, producing clutches of 5–13 neonates (197–244mm).

Siebold's Water Snake ■ *Ferania sieboldii* 78cm

DESCRIPTION Top of body is brownish grey or olive-brown, with cream or pale brown cross-bars, these sometimes fused on vertebral region, and widening on lower flanks; forehead has a dark brownish-grey or olive-brown lanceolate area; belly is cream or grey, mottled with dark green. Body is robust and sub-cylindrical; head is short, rounded and indistinct from neck; eyes are small with a vertical pupil; tail is short and compressed; dorsals are smooth; subcaudals are paired; cloacal scute is divided. **DISTRIBUTION** Myanmar. Extralimitally: Bangladesh, Nepal, India. **HABITS AND HABITAT** Rivers and water bodies within river drainage basins. Aquatic, in streams and swamps. Diet comprises frogs and fish. Ovoviviparous, producing 5–7 neonates (182mm).

Crab-eating Mangrove Snake ■ *Fordonia leucobalia* 95cm
(Bahasa Malaysia: Ular Air Ketam. Thai: Ngu Pla Hua Tao)

DESCRIPTION Top of body is variable, and can be dark grey or brown with light spots, or light grey, yellow or orange with dark or light spots; labials are yellowish cream; belly is pale cream, sometimes with small dark spots. Body is robust and cylindrical; head is short, wide and indistinct from neck; forehead scales are large and distinct; snout is rounded; lower jaw is short; eyes are small with a rounded pupil; tail is short, with an acute tip; dorsals are smooth; subcaudals are paired; cloacal scute is divided. **DISTRIBUTION** Myanmar, Thailand, Peninsular Malaysia, Singapore, Sumatra, Borneo, Java. Extralimitally: Sunderbans, Andaman and Nicobar islands, Vietnam, Lesser Sundas, Philippines, New Guinea, N Australia. **HABITS AND HABITAT** Tidal rivers; sometimes mangrove forests. Aquatic, sheltering in crab burrows. Diet comprises crabs, fish and mud lobsters. Ovoviviparous, producing 2–17 neonates (176–196mm).

Glossy Marsh Snake ■ *Gerarda prevostiana* 53cm
(Thai: Ngu Pla Ta Maeo)

DESCRIPTION Top of body is plain grey, greyish green or brown; a cream or yellow stripe extends from labials to tail tip; labials are edged with dark grey or olive; belly is grey or brownish cream with white edges, or white with grey edges; subcaudals are dark grey. Body is slender and cylindrical; head is slightly distinct from neck; eyes are small and situated

dorsally, with a vertical pupil; tail is short, with an acute tip; dorsals are smooth; cloacal scute is divided. DISTRIBUTION Myanmar, Thailand, Peninsular Malaysia, Singapore. Extralimitally: India, Bangladesh, Sri Lanka. HABITS AND HABITAT Mangrove swamps and estuaries. Diet comprises soft-shelled crabs, fish and shrimps. Ovoviviparous (numbers and size of neonates unknown).

Puff-faced Water Snake ■ *Homalopsis buccata* 140cm
(Bahasa Malaysia: Ular Air Tembam, Ular Kadut. Bahasa Indonesia: Ular Air Belalang. Cantonese Chinese: Sui Seh. Thai: Ngu Hua-kra-lok, Ngu Leuam-ao)

DESCRIPTION Top of body is variable, and can be greyish, dark brown or black with 19–51 narrow, black-edged yellow cross-bars; belly is cream with black spots. Body is robust and flattened dorso-ventrally; head is large and distinct from neck; snout is squarish; eye and nostrils are directed upward; eyes are small with a vertically oval pupil; tail is short,

with an acute tip; dorsals are keeled; cloacal scute is divided. DISTRIBUTION Southern, Thailand, Peninsular Malaysia, Singapore, Sumatra, Pulau Belitung, Borneo, Java. HABITS AND HABITAT Sluggish and stagnant waterways, including peat swamps, ponds, marshes, rice fields and coastal areas. Diet comprises fish, crustaceans and frogs. Ovoviviparous, producing 2–37 neonates (192–260mm).

Marquis Doria's Water Snake ■ *Homalophis doriae* 79.6cm

DESCRIPTION Top of body is reddish brown to greyish brown; labials are blotched with grey or cream; belly is yellow, orange, cream or red, some individuals with dark blotches, these darker posteriorly. Body is robust and sub-cylindrical; head is small, depressed and slightly distinct from neck; forehead scales are fragmented; eyes are small with a rounded or slightly oval pupil; tail is short; dorsals are smooth; subcaudals are paired; cloacal scute is divided.
DISTRIBUTION Borneo.
HABITS AND HABITAT Small freshwater streams and swamps, including peat lakes (<500m asl). Aquatic. Diet comprises fish. Reproductive habits are unstudied.

Gyi's Water Snake ■ *Homalophis gyii* 76.2cm

DESCRIPTION Top of body is iridescent greyish brown, except for scale rows 1–4, which are reddish brown, each dorsal with a red spot (colour changeable to nearly white when individuals are kept in the dark); supralabials from anterior to orbit are greyish black; posterior supralabials are reddish brown; a black stripe extends from nape to angle of jaws; belly is reddish brown. Body is robust and sub-cylindrical; head is distinct from neck; eyes are small with a vertical pupil; dorsals are smooth; subcaudals are paired; cloacal scute is divided. **DISTRIBUTION** Borneo. **HABITS AND HABITAT** Known from a single locality – Putussibau, near Sungai Kapuas, within swamp forests (*c.* 50m asl). Diet and reproductive biology are unstudied.

Cambodian Puff-faced Water Snake ▪ *Homalopsis mereljcoxi* 97.3cm

DESCRIPTION Top of body with 20–28 pale bands bordered in black, separated by dark brown blotches; bands cross vertebral region at midbody; first three dorsal rows uniform yellow; lack the pattern; belly is white or yellow-cream with dark lateral spot on outer edge of some scales; undersurface of tail mottled with black and cream. **DISTRIBUTION** Thailand. Extralimitally: Cambodia and Vietnam. **HABITS AND HABITAT** Inhabits canals, streams, rivers, reservoirs, ditches and ponds. Diet comprises fish. Ovoviviparous, producing up to 17 young.

Martaban Puff-faced Water Snake ▪ *Homalopsis semizonatus* 73cm

DESCRIPTION Top of body pale yellowish-brown, with 36 dark cross-bars, that are paler within; six or more dorsal rows between ventral scales and bottom of brown blotches; triangular black snout spot and eye-streak; belly with two irregular rows of dark spots. **DISTRIBUTION** Myanmar (Gulf of Martaban). **HABITS AND HABITAT** Hill streams in evergreen as well as deciduous forests, and also ponds and marshes, and water bodies within cities. Diet includes frogs. Ovoviviparous; clutch size unknown.

Grey Water Snake ■ *Hypsiscopus plumbea* 71.3cm
(Bahasa Malaysia: Ular Air Sawah. Bahasa Indonesia: Ular Lumpur. Thai: Ngu Pling)

DESCRIPTION Top of body is grey or greyish olive, each scale edged with dark brown or black; supralabials and belly are cream or yellow, the latter with black spots; anal and subcaudals have a dark grey median line. Body is robust and sub-cylindrical; head is short, rounded and indistinct from neck; eyes are small with a rounded pupil; tail is short and compressed; dorsals are smooth; subcaudals are paired; cloacal scute is divided. **DISTRIBUTION** Myanmar, Thailand, Peninsular Malaysia, Sumatra, Pulau Belitung, Borneo, Java, Bali. Extralimitally: Laos, Cambodia, Vietnam, China, Sulawesi, Nicobar Archipelago. **HABITS AND HABITAT** Wetlands, such as swamps, marshes, streams, ditches and paddy fields; occasionally reported from brackish waters. Diet comprises crustaceans, frogs and their eggs, tadpoles and fish. Ovoviviparous, producing 2–30 neonates (size unknown).

Pahang Water Snake ■ *Kualatahan pahangensis* 26.5cm

DESCRIPTION Top of body is grey-brown with small, dark spots; a pale yellow stripe on flanks, bordered by a dark zigzag line, covers first 4 dorso-lateral scale rows anteriorly and 3 rows posteriorly; has wide white stripes on sides of head, supralabials and rostral; belly

is white. Body is robust and sub-cylindrical; head is distinct from neck; eyes are small with a vertical pupil; dorsals are smooth; subcaudals are paired; cloacal scute is divided. **DISTRIBUTION** Peninsular Malaysia (known from Pahang and Terengganu). **HABITS AND HABITAT** Hill dipterocarp forests (152–305m asl). Aquatic, in streams. Diet and reproductive habits are unstudied.

Bocourt's Water Snake ■ *Subsessor bocourti* 110cm

DESCRIPTION Back of body reddish-brown or dark brown, with transverse, black-edged, yellowish-brown bars, which are narrow on flanks to meet ventrals; head greenish-brown; supralabials cream with black bars on sutures; belly yellow. **DISTRIBUTION**

Thailand (Beung Boraphet, Nakhon Sawan Province, Bangkok, Phra Nakhon Si Ayutthaya Province, Nong Thung Thong, Surat Thani Province and Thale Noi, Phattalung Province), and northern Peninsular Malaysia. Extralimitally: Cambodia and Vietnam. **HABITS AND HABITAT** Stagnant or slow-moving freshwater bodies in the lowlands and low hills, up to *c.* 200m asl. Diet comprises fish. Ovoviviparous, producing 15–20 neonates, measuring 22cm.

White-spotted Water Snake ■ *Sumatranus albomaculata* 65cm

DESCRIPTION Top of body olive brown or black, with small yellow or orange spots; yellow nuchal collar and 1 or 2 indistinct yellow cross-bars on forehead; belly with yellow,

olive brown or black pattern. Body robust, subcylindrical; scales smooth; last 2 or 3 upper labials horizontally divided; large internasal makes narrow contact with elongated loreal; 5 of 6 lower labials contact anterior chin shields. **DISTRIBUTION** Sumatra, as well as adjacent islands of Pulau Nias, Pulau Simeuleu and Pulau Sibigo. **HABITS AND HABITAT** Stagnant and slow-moving freshwater bodies. Diet comprises fish. Reproductive habits unstudied, except for being ovoviviparous.

Blunt-headed Slug Snake ▪ *Aplopeltura boa* 85cm
(Thai: Ngu Kin Tak Hua Nok, Ngu Boa)

DESCRIPTION Top of body is brown to greyish brown, typically with dark-edged saddle-like markings; flanks often have large white spots; forehead is dark brown; labials are cream; has a cream patch with a dark sub-triangular area under eye; belly is brown to dark grey. Body is slender and laterally compressed; head is short, rounded and distinct from neck; snout is short; eyes are large with a vertical pupil; tail is a third of body length; dorsals are smooth; cloacal scute is entire.

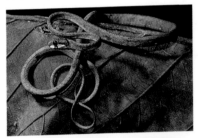

DISTRIBUTION Myanmar, Thailand, Peninsular Malaysia, Sumatra, Pulau Nias, Natuna Archipelago, Borneo, Java. Extralimitally: Philippines. **HABITS AND HABITAT** Lowland and sub-montane forests (<1,500m asl). Arboreal, in low vegetation of bushes and undergrowth. Diet includes slugs, snails and lizards. Oviparous, clutches comprising 4–8 eggs (18–23 × 10–13.5mm).

Malayan Slug Snake ■ *Asthenodipsas borneensis* 41cm

DESCRIPTION Top of body is light brown, with irregular, brownish-grey cross-bars, 2–4 scales wide; forehead is white to greyish-brown; dark brown or black band on neck and forebody encircles body except for a narrow gap on vertebral region; belly is yellowish cream, suffused with brown laterally. Body is robust and laterally compressed; head is short, rounded and distinct from neck; snout is short; eyes are small with a vertical pupil; tail is

short; vertebral scale row is enlarged and vertebrals have a distinct keel; dorsals are weakly keeled; subcaudals are paired; cloacal scute is entire. **DISTRIBUTION** Borneo. **HABITS AND HABITAT** Lowland dipterocarp forests (<1,000m asl). Semi-arboreal and terrestrial, on low vegetation and the forest floor. Diet comprises snails and slugs. Reproductive habits are unstudied.

Smooth Slug Snake ■ *Asthenodipsas laevis* 60cm
(Bahasa Malaysia: Ular Kapak Rimau. Thai: Ngu Kin Tak Gled Reab)

DESCRIPTION Top of body is mid-brown to dark brown, with numerous dark vertical bars that extend to belly; forehead is darker than dorsum and lacks lines; throat is brown; belly is cream or pale yellow, edges of each scale with a dark spot. Body is slender and laterally compressed; head is short, rounded and distinct from neck; snout is short; eyes are large with a vertical pupil; tail is short; vertebrals are slightly enlarged and have a distinct keel; dorsals

are smooth; subcaudals are paired; cloacal scute is entire. **DISTRIBUTION** Thailand, Peninsular Malaysia, Sumatra, Pulau Bangka, Natuna and Mentawai archipelagos, Borneo, Java. **HABITS AND HABITAT** Lowland forests (<1,150m asl). Semi-arboreal and terrestrial, in low vegetation and on the forest floor. Diet comprises slugs and snails. Reproductive biology is unstudied.

Mountain Slug-eating Snake
■ *Asthenodipsas vertebralis* 77.1cm

DESCRIPTION Top of body is reddish brown or dark brown, with small dark brown spots and indistinct dark cross-bars; has an interrupted yellow vertebral stripe; belly is yellow with brown spots laterally. Body is slender and laterally compressed; head is short and rounded, and distinct from neck; snout is short; eyes are large with a vertical pupil; tail is short; dorsals are smooth; cloacal scute is entire. **DISTRIBUTION** Peninsular Malaysia. **HABITS AND HABITAT** Sub-montane and montane forests (1,585–2,012m asl). Arboreal, on low vegetation. Diet comprises snails and slugs. Reproductive biology is unstudied.

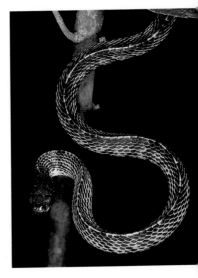

Keeled Slug-eating Snake ■ *Pareas carinatus* 60cm
(Thai: Ngu Kin Tak Gled San)

DESCRIPTION Top of body is olive-brown, yellow or reddish brown, with indistinct yellowish-white transverse bars at anterior; a dark streak is present along each eye; belly is pale brown to yellow. Body is slender and laterally compressed; head is short, rounded and distinct from neck; snout is short; eyes are large with a vertical pupil; tail is short; dorsals are enlarged and weakly keeled on 2 median rows; subcaudals are paired; cloacal scute is entire. **DISTRIBUTION** Myanmar, Thailand, Peninsular Malaysia, Sumatra, Borneo, Java, Bali. Extralimitally: Laos, Cambodia, Vietnam, S China. **HABITS AND HABITAT** Lowland dipterocarp and sub-montane forests (550–1,300m asl). Arboreal, on low vegetation. Diet comprises molluscs. Oviparous, clutches comprising 3–8 eggs (19–25 × 9–12mm).

Hampton's Slug-eating Snake ▪ *Pareas hamptoni* 70.5cm
(Thai: Ngu Kin Tak Lai Kwan)

DESCRIPTION Top of body is light brown, with vertical black cross-bars; forehead has with dense black spots; belly is yellow with brown spots. Body is slender and laterally

compressed; head is short, rounded and distinct from neck; snout is short; eyes are large with a vertical pupil; tail is short; dorsals are smooth; 1–3 vertebral rows are enlarged; subcaudals are paired; cloacal scute is entire. **DISTRIBUTION** Myanmar, Thailand. Extralimitally: Laos, Vietnam, E China. **HABITS AND HABITAT** Mid-hill to sub-montane forests. Arboreal, in vegetation near forest streams. Diet and reproductive habits are unstudied.

White-spotted Slug-eating Snake ▪ *Pareas margaritophorus* 45cm
(Thai: Ngu Kin Tak Chut Khao)

DESCRIPTION Top of body is light or dark grey, with irregular black cross-bars bordered with white; labials are cream with black mottling; a pale nuchal collar is sometimes present; belly is cream with irregular black speckles. Body is slender and not laterally compressed; head is short, rounded and distinct from neck; snout is short; eyes are large with a vertical pupil; tail is short; vertebral scale rows are enlarged; dorsals are smooth; subcaudals are paired; cloacal scute is entire. **DISTRIBUTION** Peninsular Malaysia, Myanmar, Thailand. Extralimitally: Laos, Cambodia, Vietnam, E China, E India. **HABITS AND HABITAT** Evergreen forests of bamboo and sub-montane forests (340–1,524m asl). Arboreal, on shrubs. Diet comprises snails and slugs. Oviparous, clutches comprising 6 eggs (size unknown).

Montane Slug-eating Snake
■ *Pareas monticola* 61cm

DESCRIPTION Top of body is mid-brown, with vertical blackish-brown bars on flanks; a black post-ocular stripe extends to nape; forehead is brown with dense black spots; belly is yellow with brown spots. Body is slender and laterally compressed; head is short, rounded and distinct from neck; snout is short; eyes are large with a vertical pupil; tail is short; vertebral rows are enlarged; dorsals are weakly keeled; subcaudals are paired; cloacal scute is entire. **DISTRIBUTION** Myanmar. Extralimitally: Vietnam, E India. **HABITS AND HABITAT** Evergreen hill forests (<c. 1,800m). Arboreal, on low vegetation. Diet includes snails and slugs. Oviparous, laying up to 8 eggs (size unknown).

Barred Slug-eating Snake ■ *Pareas nuchalis* 71.5cm

DESCRIPTION Top of body is mid-brown to tan, anterior with narrow bands that are fused vertebrally and separated by 2–5 scales; a dark post-ocular stripe runs to angle of jaws, extending to nuchal region and fusing with a dark, irregular V-shaped nuchal patch; belly is yellowish cream. Body is slender and laterally compressed; head is short, rounded and distinct from neck; snout is short; eyes are large with a vertical pupil; tail is short; dorsals are weakly keeled; vertebrals are enlarged; subcaudals are paired; cloacal scute is entire. **DISTRIBUTION** Borneo. **HABITS AND HABITAT** Lowland dipterocarp forests (up 850m asl). Arboreal, on shrubs and other low vegetation. Diet comprises snails and slugs. Reproductive habits are unstudied.

VIPERIDAE – VIPERS AND PIT VIPERS
Members of this familiar family include vipers and pit vipers, a group of generally slow-moving species with haemotoxic venom. They are identifiable by their broad triangular heads, hinged fangs that fold back against the palate of mouth when not in use, and elongate ectopterygoid bone that functions as a lever to move the fang. Additionally, their dorsals tend to be heavily keeled. In pit vipers, specialised pit organs are present between the nostril and the eye; these contain infrared receptors that permit the detection of prey through thermal cues. Members of Viperidae are encountered in habitats as diverse as scrubland, mangrove swamps, tropical rainforests and temperate forests. Their diet primarily comprises birds and mammals, although insects and other large invertebrates may also be eaten. They are mainly ovoviviparous, with oviparity shown in several species, and have a wide-ranging distribution.

Fea's Viper ▪ *Azemiops feae* 92.5cm

DESCRIPTION Top of body is blackish brown with grey-edged scales, and with 14–15 narrow white or pink cross-bars that may be interrupted mid-dorsally or alternate laterally; forehead is yellow with a pair of dark brown to black stripes extending from pre-frontals to end of neck. Body is robust and cylindrical; sensory pits are absent; head is flattened and distinct from neck; forehead is covered with large, symmetrical shields; nostrils are large; eyes are small with a vertical pupil; dorsals are smooth; subcaudals are mostly paired; cloacal scute is entire. **DISTRIBUTION** Myanmar. The records from north-eastern Vietnam and eastern China refer to the recently-described *Azemiops kharini*. **HABITS**

AND HABITAT Sub-montane and montane forests, and paddy fields near villages (700–2,000m asl). Terrestrial, in forests of bamboo and tree ferns. Diet includes rodents and shrews. Oviparous, clutches comprising 5 eggs (size unknown). **VENOM** Potentially lethal envenoming, insufficient clinical reports. No antivenom available.

Malayan Pit Viper ■ *Calloselasma rhodostoma* 145cm
(Bahasa Malaysia: Ular Kapak Daun. Bahasa Indonesia: Ular Donda. Thai: Ngu Maaw Sao)

DESCRIPTION In adults, top of body is reddish brown or purplish brown; flanks are paler and speckled with dark brown; a series of 19–31 dark brown sub-triangular marks on each side, their tips meeting or alternating; a dark post-ocular stripe, scalloped ventrally, extends to angle of jaws; belly is pinkish cream, mottled with brown. Body is robust; head is distinct from neck; snout is acute and slightly upturned; tail is thin and short; vertebral ridge is distinct; eyes with a vertical pupil; dorsals are smooth; cloacal scute is entire. **DISTRIBUTION** Thailand, Laos, Cambodia, Vietnam, Peninsular Malaysia, Java, Karimun Jawa and Kangean archipelagos. Extralimitally: Laos, Cambodia, Vietnam. **HABITS AND HABITAT** Dry lowland forests and plantations, reaching sub-montane limits (<1,524m asl).

Terrestrial, in dense undergrowth and rocky biotopes. Diet comprises small rodents and birds. Oviparous, producing 13–30 eggs (diameter 20–30mm). **VENOM** Fibrinogenases and haemorrhagins; linked to severe envenoming that is potentially lethal. Malayan Pit Viper Antivenin (Thai Red Cross Society); Antivenin Polyvalent (Equine) (P.T. Bio Farma, Persero, Bandung); Malayan Pit Viper Antivenom (Thai Government Pharmaceutical Organisation).

White-lipped Pit Viper ■ *Cryptelytrops albolabris* 104cm
(Thai: Ngu Kieo Hang-mai Tong Loeng)

DESCRIPTION Top of body and forehead are bright green or lime-green; sides of head below eye, including labials, are yellow, white or pale green, lighter than rest of head; tail is reddish brown; iris is yellow; males have a white stripe on 1st row of dorsal scales, females are unstriped. Body is robust; head is relatively long and distinct from neck; forehead scales are small; eyes are large with a vertical pupil; tail is short and prehensile; dorsals are smooth or weakly keeled. **DISTRIBUTION** Myanmar, Thailand, Peninsular Malaysia. Extralimitally: Bangladesh, Nepal, E India, Nicobar Islands, Laos, Cambodia, Vietnam, China, Lesser Sundas, Sulawesi. **HABITS AND HABITAT** Deciduous, subtropical and temperate forests, near streams; also near agricultural

fields (60–3,050m asl). Arboreal, on trees and saplings, and occasionally on ground. Diet comprises rodents, birds, lizards and frogs. Ovoviviparous, producing 7–17 neonates (120–195mm). **VENOM** Fibrinogenases and possibly haemorrhagins. Severe envenoming that is potentially lethal. Polyvalent Anti Snake Venom Bivalent Antivenin Pit Viper, Trimeresurus antivenin (National Institute of Preventative Medicine, Taiwan).

Spot-tailed Pit Viper ■ *Cryptelytrops erythrurus* 105cm

DESCRIPTION Top of body and head are dark green; labials are yellow; males have a white dorso-lateral stripe from posterior edge of eye to tail; this is indistinct or absent in females; belly is pale green or yellow; tip of tail is usually spotted or mottled with brown. Body is slender; head is relatively long, narrower than neck; small forehead scales are present; temporals are small and strongly keeled; eyes are large with a vertical pupil; tail is short and prehensile; dorsals are strongly keeled; subcaudals are paired; cloacal scute is entire. **DISTRIBUTION** Myanmar. Extralimitally: Bangladesh, Nepal, E India. **HABITS**

AND HABITAT Lowland rainforests and moist deciduous forest (<200m asl). Arboreal, on trees; also on ground, near streams. Diet comprises rodents, birds and lizards. Ovoviviparous, producing 10 neonates (231–260mm).
VENOM Procoagulants, anticoagulants and haemorrhagins possibly present. Bites unstudied but with potentially lethal envenoming. Polyvalent Anti Snake Venom Serum (Central Research Institute (Kasauli); SII Bivalent Antisnake Venom Serum (lyophilized) (Serum Institute of India Ltd).

Lesser Sunda White-lipped Pit Viper ■ *Cryptelytrops insularis* 93cm
(Bahasa Indonesia: Ular Mati Ekor)

DESCRIPTION Top of body is typically bright green (sometimes olive or blue), with dark transverse bands; supralabials are yellow to greenish white; tail has a rust-coloured streak; belly is greenish yellow, greenish white or light blue; iris is brown or red. Body is slender; head is distinct from neck; forehead scales are granular; temporals have a short keel; eyes are large with a vertical pupil; tail is short and prehensile; dorsals are keeled; cloacal scute

is entire. **DISTRIBUTION** Bali. Extralimitally: Komodo, Flores, Kisar, Lombok, Sumba, Sumbawa, Timor. **HABITS AND HABITAT** Lowland forests (<880m asl). Arboreal, on trees up to 15m. Diet comprises skinks, geckos, rodents and frogs. Ovoviviparous, producing 11–17 neonates (140–195mm). **VENOM** Fibrinogenases and possibly haemorrhagins possibly present. Severe, even lethal envenoming possible. No antivenom raised.

Kanburi Pit Viper ▪ *Cryptelytrops kanburiensis* 66.7cm
(Thai: Ngu Hang-ham Kan)

DESCRIPTION Top of body is olive or greyish green; has dull brown or orange-brown blotches on head; has white or blue spots on labials and on 1st dorsal row; has a broad, dark post-ocular stripe; belly is cream; some ventrals have brown edges, especially in males; iris is light orange with brown speckles. Body is slender; head is distinct from neck; supra-oculars are irregular and indented; temporals are keeled; eyes are large with a vertical pupil; tail is short and prehensile; dorsals are keeled, except for lowest row; cloacal scute is entire. **DISTRIBUTION** Thailand. **HABITS AND HABITAT** Evergreen forests in limestone hills of Kanchanaburi Province, especially those dominated by bamboo. Terrestrial. Diet and reproductive habits are unstudied. **VENOM** Procoagulants, anticoagulants, haemorrhagins, nephrotoxins and necrotoxins possibly present. Bites unstudied and likely to cause severe envenomation. Green Pit Viper Antivenin (Thai Red Cross Society).

Large-eyed Pit Viper ▪ *Cryptelytrops macrops* 71cm

DESCRIPTION Dorsum pale green, sometimes with a pale blue lateral stripe along first dorsal scale row; labials pale bluish-green; forehead pale green, juveniles and males with a white postocular stripe; tail reddish-brown; chin and gular region bluish-white; belly pale blue; subcaudals pale blue; iris golden yellow in females; reddish-orange in males. **DISTRIBUTION** Central and south-eastern Thailand. Extralimitally: Cambodia, Laos and southern Vietnam. **HABITS AND HABITAT** Evergreen hill forests, at elevations between 200–500m asl. Diet comprises birds, small mammals, lizards and tree frogs. Ovoviviparous, producing 6–12 live young.

Mangrove Pit Viper ▪ *Cryptelytrops purpureomaculatus* 104cm
(Bahasa Malaysia: Ular Kapak Bakau. Thai: Ngu Pang-ka)

DESCRIPTION Top of body is variable in colour, ranging from olive to grey and brownish purple, and with darker mottling; forehead is olive with intense brown mottling; belly is cream-green or brown, unpatterned or spotted with brown; a pale line on scale row bordering ventrals is sometimes present. Body is slender; head is distinct from neck;

forehead scales are small, tuberculate or granular; temporals are keeled; eyes are large with a vertical pupil; tail is short and prehensile; dorsals are keeled; subcaudals are paired; cloacal scute is entire. **DISTRIBUTION** Myanmar, Thailand, Peninsular Malaysia, Singapore, Sumatra. **HABITS AND HABITAT** Mangrove and coastal swamp forests. Arboreal, on shrubs and trees. Diet includes small mammals. Reproductive habits are unstudied. **VENOM** Procoagulants and anticoagulants possibly present; haemorrhagins present. Severe, even lethal envenoming possible. Green Pit Viper Antivenin (Thai Red Cross Society).

Eastern Russell's Viper ▪ *Daboia siamensis* 185cm
(Bahasa Indonesia: Ular Viper Russell. Thai: Ngu Meaw Sao)

DESCRIPTION Top of body is light brown or greyish brown with 5–7 rows of large brown blotches, a dark brown one along midline and a blackish-brown or black one on flanks; belly is yellowish cream with numerous small crescent marks on ventrals. Body is robust; head is distinct from neck; forehead has small scales; nostrils are enlarged, in large nasal shields; eyes are small with a vertical pupil; tail is short; all mid-body scales, except outermost row, are strongly keeled; subcaudals are paired; cloacal scute is entire. **DISTRIBUTION** Myanmar, Thailand. Extralimitally: Cambodia, Java, E China, Lesser Sundas. **HABITS AND HABITAT** Lowland grasslands, scrub forests and other open forests, and agricultural fields (<2,756m

asl). Terrestrial. Diet comprises rodents, crabs, frogs, lizards and birds. Ovoviviparous. **VENOM** Mixture of procoagulants; likelihood of severe envenoming and high lethality. Russell's Viper Antivenin (Thai Red Cross Society); Russell's Viper Antivenom (Thai Government Pharmaceutical Organisation); Bivalent (Pharmaceutical Industries Corporation, Yangon); Anti-Viper, Russell's Viper (Pharmaceutical Industries Corporation, Yangon).

Kinabalu Brown Pit Viper ■ *Garthius chaseni* 69cm

DESCRIPTION Top of body is dark or reddish brown, marked with irregular dark brown blotches in paired rows anteriorly, joining to form bands towards posterior of body; smaller dark blotches are present on lower flanks; a dark post-ocular stripe, edged ventrally by a line of yellow scales, extends to neck; belly is yellow with large grey areas. Body is robust; head is triangular, broad and flattened, distinct from neck; snout is blunt; eyes are small with a vertical pupil; tail is not prehensile; dorsals are strongly keeled anteriorly, weakly keeled posteriorly; subcaudals are paired; cloacal scute is entire. **DISTRIBUTION** Endemic to Gunung Kinabalu and the Crocker Range, Sabah, N Borneo. **HABITS AND HABITAT** Sub-montane forests (915–1,550m asl). On leaf litter. Diet unknown but probably includes small birds and mammals. Ovoviviparous (neonate numbers and size unknown). **VENOM** Mixture of procoagulants and probably anticoagulants, and venom haemorrhagins. Bites with potentially lethal envenoming. No antivenom raised.

Malayan Brown Pit Viper ■ *Ovophis convictus* 71cm

DESCRIPTION Top of body is mid-brown or brownish yellow, with sub-rectangular, dark brown or black blotches from nape to mid-tail; forehead is greyish black or dark brown; a yellow post-ocular stripe extends up to neck; belly is cream with brown mottling. Body is robust; head is triangular, broad and flattened, and distinct from neck; eyes are small with a vertical pupil; tail is not prehensile; dorsals are keeled; subcaudals are paired; cloacal scute is entire. **DISTRIBUTION** Thailand, Peninsular Malaysia, Singapore. Extralimitally: Cambodia, Vietnam. **HABITS AND HABITAT** Sub-montane and montane forests (<*c.* 1,737m asl). Terrestrial. Diet comprises small mammals. Ovoviviparous, producing 21 neonates (size unknown). **VENOM** Probably a mixture of procoagulants and probably anticoagulants and venom haemorrhagins. Bites with potentially lethal envenoming. Green Pit Viper Antivenin (Thai Red Cross Society); Bivalent Antivenin Pit Viper, Trimeresurus antivenin (National Institute of Preventative Medicine, Taiwan).

Blotched Pit Viper ▪ *Ovophis monticola* 125cm
(Thai: Ngu Hang-han Phu-khao)

DESCRIPTION Top of body is brownish grey or yellowish pink with squarish, dark brown blotches; has a lateral series of smaller, dark brown spots; forehead is darker, with a pale post-ocular stripe reaching neck; belly is dusted with dark brown. Body is robust; head is triangular, broad and flattened, and distinct from neck; supra-oculars are large and entire; eyes are small with a vertical pupil; tail is not prehensile; dorsals are either smooth or weakly keeled; subcaudals are either single or paired; cloacal scute is entire. **DISTRIBUTION** Myanmar, Thailand. Extralimitally: Laos, Vietnam, E India, Nepal, Bangladesh, China. **HABITS AND HABITAT** Montane and sub-montane evergreen forests

(700–1,200m asl). Terrestrial. Diet includes rodents, birds and their eggs, frogs and lizards. Ovoviviparous, laying 5–18 eggs (diameter 18–20mm). **VENOM** Mixture of procoagulants and probably anticoagulants and venom haemorrhagins. Bites with potentially lethal envenoming. Green Pit Viper Antivenin (Thai Red Cross Society); Bivalent Antivenin Pit Viper, Trimeresurus antivenin (National Institute of Preventative Medicine, Taiwan).

Hagen's Green Pit Viper ▪ *Parias hageni* 116cm
(Thai: Ngu Kieo Hang-mi Hagen)

DESCRIPTION Top of body is bright green; forehead and body scales are not edged with black; a pale line, bordered ventrally by a dark line or series of dark spots, extends along 1st and 2nd dorsal rows; lip scales, chin and throat regions are pale green; belly is pale green, ventrals sometimes with dark edges. Body is robust; head is distinct from neck; forehead scales are smooth; eyes are small with a vertical pupil; tail is prehensile; dorsals are weakly keeled; cloacal scute is entire. **DISTRIBUTION** Thailand, Peninsular Malaysia, Singapore,

possibly Mentawai Archipelago. **HABITS AND HABITAT** Lowland and mid-hill forests (<600m asl). Arboreal, on trees and saplings. Diet comprises birds and mammals. Reproductive habits are unstudied. **VENOM** Procoagulants, anticoagulants and haemorrhagins possibly present. Bites unstudied and likely to cause severe envenoming. Green Pit Viper Antivenin (Thai Red Cross Society).

Kinabalu Green Pit Viper ■ *Parias malcolmi* 133cm

DESCRIPTION Top of body is black, each scale with a sub-triangular green pattern, creating indistinct bands; forehead is black with green-centred scales; juveniles have a white stripe along flanks, absent in adults; belly is pale green, with green ventrals that have a dark posterior edge and pale green anterior edge; tail has parallel red dots within dark green scales; subcaudals are light green, edged with black. Body is robust; head is narrow and distinct from neck; forehead scales are smooth; eyes are small with a vertical pupil; tail is prehensile; dorsals are weakly keeled; cloacal scute is entire. **DISTRIBUTION** Gunung Kinabalu, Sabah, N Borneo. **HABITS AND HABITAT** Sub-montane and montane forests (1,000–1,700m asl). Arboreal, in oak forests; also secondary vegetation. Diet unknown but presumably comprises rodents. Reproductive habits are unstudied. **VENOM** Procoagulants, anticoagulants and haemorrhagins possibly present. Bites unstudied and likely to cause severe envenoming. No antivenom raised.

Sumatran Pit Viper ■ *Parias sumatranus* 135.5cm
(Bahasa Malaysia/Indonesia: Ular Kapak. Iban: Ular Beliung, Ular Engkaradau. Thai: Ngu Kieo Hang-mi Sumatra)

DESCRIPTION Top of body is green with black cross-bars, these especially noticeable in adults; forehead scales are edged with black; supralabials are light blue; tail is reddish brown, brighter in juveniles; belly is yellowish green. Body is robust; head is distinct from neck; forehead scales are smooth; internasals are present; eyes are small with a vertical pupil; tail is prehensile; dorsals are weakly keeled; cloacal scute is entire. **DISTRIBUTION** Thailand, Peninsular Malaysia, Singapore, Sumatra, Mentawai Archipelago, Pulau Belitung, Borneo, possibly Palawan. **HABITS AND HABITAT** Lowlands and mid-hill forests (<1,000m asl). Arboreal, on low vegetation. Diet comprises mammals, birds and frogs. Reproductive habits are unstudied. **VENOM** Anticoagulants and haemorrhagins present; procoagulants possibly present. Bites unstudied and likely to cause severe envenoming. Green Pit Viper Antivenin (Thai Red Cross Society).

Pulau Tioman Pit Viper
▪ *Popeia buniana* 78.3cm

DESCRIPTION Top of body is pale turquoise, with 81 transverse, zigzagging maroon bands on body and 39 brownish bands on tail; iris centre is copper, outer edge is turquoise; males have a maroon post-orbital stripe and bicoloured ventro-lateral stripe; females have a white ventro-lateral stripe on flanks and tail, bordered ventrally by a red stripe, fading caudally. Body is long and slender; head is triangular and elongate, distinct from neck; snout is pointed; rostral is large and triangular; eyes are small with a vertical pupil; cloacal scute is entire. **DISTRIBUTION** Pulau Tioman, Peninsular Malaysia. **HABITS AND HABITAT** Lowland and hill dipterocarp forest transitions to hill dipterocarp forest (240–810m asl). Arboreal, on trees 2–10m above ground. Diet includes frogs and, possibly, lizards. Reproductive habits are unstudied. **VENOM** Venom characteristics and bite remain unstudied. Green Pit Viper Antivenin (Thai Red Cross Society).

Thai Peninsula Pit Viper ▪ *Popeia fucata* 86.8cm

DESCRIPTION Top of body is green in both sexes, with irregular rusty or reddish-brown cross-bands; males have a white, or white and reddish-brown, post-ocular streak and white dots on vertebral region, and a bright ventro-lateral stripe, orange or red ventrally and white dorsally; females lack post-ocular streak and vertebral dots, and ventro-lateral stripe is thin and white; tail is rusty or reddish brown, sometimes mottled; pupil is yellowish green, greenish gold or yellow-copper. Body is robust and cylindrical; head is triangular and distinct from neck; eyes are small with vertical pupil; tail is long; occipital and temporals

are keeled; snout is elongate and obliquely truncate; cloacal scute is single. **DISTRIBUTION** Myanmar, Thailand, Peninsular Malaysia. **HABITS AND HABITAT** Mid-hills to sub-montane and montane forests (400–1,280m asl). Arboreal, on trees. Diet includes rodents and, possibly, birds. Reproductive habits are unstudied. **VENOM** Venom characteristics and bite remain unstudied. Green Pit Viper Antivenin (Thai Red Cross Society).

Cameron Highlands Pit Viper ■ *Popeia nebularis* 100.2cm

DESCRIPTION Top of body is bright green, with a hint of blue; supralabials are bluish green; chin and throat are yellowish green; post-ocular streak and ventro-lateral stripe are typically absent; tail is rusty brown vertebrally and green laterally, with a sharp border between; iris is green or yellowish green. Body is robust and cylindrical; head is distinct from neck; eyes are small with a vertical pupil; dorsals are keeled. **DISTRIBUTION** Peninsular Malaysia. **HABITS AND HABITAT** Montane forests. Arboreal; also encountered on the ground. Diet is unknown but presumably includes rodents and birds. Reproductive habits are unstudied. **VENOM** Venom characteristics and bite remain unstudied. Green Pit Viper Antivenin (Thai Red Cross Society).

Phuket Pit Viper ■ *Popeia phuketensis* 64cm

DESCRIPTION Top of body is green, with 78 reddish-brown cross-bands; has a post-ocular streak, reddish brown in its wide upper part and white in its narrow lower part; narrow ventro-lateral stripe is white and red in males, pale green and red in females; forehead is green with maroon markings; supralabials are paler green. Body is long and thin; head is triangular and elongate; occipitals are keeled; temporals are smooth; eyes are small with a vertical pupil; dorsals are keeled; tail is long, with a prehensile tip; cloacal scute is entire. **DISTRIBUTION** Thailand. **HABITS AND HABITAT** Mature primary and secondary forest areas of Ban Bang Rong, Phuket Island. Arboreal, mostly on shrubs and undergrowth up to 3m above ground. Diet in the wild is unknown; in captivity, accepts frogs, rodents, bats and lizards. Ovoviviparous, producing 8–9 neonates (size unknown). **VENOM** Venom characteristics and bite remain unstudied. Green Pit Viper Antivenin (Thai Red Cross Society).

Pope's Pit Viper ▪ *Popeia popeiorum* 105cm
(Thai: Ngu Kieo Hang-mai Tong Kieo)

DESCRIPTION Top of body is bright green or bluish green, and lacks dark cross-bars; post-ocular streak in males is narrow and white ventrally and broad and red dorsally, and in females narrow and white or absent; ventro-lateral stripe in males is bright red ventrally

and white dorsally, and in females white; belly is light green; tail is brown, mottled with green laterally; tail tip is reddish brown; iris is red in adults, yellow in juveniles. Body is slender in males, robust in females; head is distinct from neck; tail is prehensile; forehead scales are smooth; eyes are small with a vertical pupil; dorsals are smooth or weakly keeled; subcaudals are paired; cloacal scute is entire. **DISTRIBUTION** Myanmar, Thailand. Extralimitally: E India, Laos. **HABITS AND HABITAT** Montane and sub-montane forests. Arboreal, on low vegetation of shrubs and dense undergrowth. Diet comprises birds, frogs, lizards and mammals. Ovoviviparous, producing 10–15 neonates (120–200mm). **VENOM** Venom characteristics and bite remain unstudied. Green Pit Viper Antivenin (Thai Red Cross Society).

Sabah Green Pit Viper
▪ *Popeia sabahi* 81cm

DESCRIPTION Top of body is bright green, without cross-bars or a post-ocular streak; ventro-lateral stripe is red or rusty-red in males, and white or yellow in females; iris is red or orange in adults, and orange or yellowish green in juveniles. Body is slender in males, robust in females; head is triangular and distinct from neck; occipitals and temporals are smooth or weakly keeled; eyes are small with a vertical pupil; dorsals are keeled; subcaudals are keeled; cloacal scute is entire. **DISTRIBUTION** N Borneo, including Gunung Kinabalu, Crocker Range, Gunung Lumaku, Mendolong, Muruk Miau, Gunung Dulit, Gunung Gading, Gunung Penrissen, Gunung Semedoem. **HABITS AND HABITAT** Sub-montane forests (1,000–1,150m asl). Arboreal, on low vegetation of shrubs and branches. Diet and reproductive biology are unstudied. **VENOM** Venom characteristics and bite remain unstudied. Green Pit Viper Antivenin (Thai Red Cross Society).

Jerdon's Pit Viper ■ *Protobothrops jerdonii* 99cm

DESCRIPTION Top of body is greenish yellow or olive; a series of black-edged reddish-brown blotches is present on dorsum; forehead is black, with fine yellow lines. Body is slender; head is distinct from neck; snout is elongate; forehead scales are reduced; forehead, apart from internasals and supra-oculars, is covered with small, smooth scales; temporals are smooth; eyes are small with a vertical pupil; dorsals are strongly keeled; subcaudals are paired; cloacal scute is entire. **DISTRIBUTION** Myanmar. Extralimitally: Cambodia, Vietnam, E India, China. **HABITS AND HABITAT** Montane and temperate forests, including secondary forests

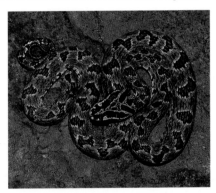

(1,200–2,300m asl). Arboreal as well as terrestrial. Diet is unknown but presumably comprises rodents and birds. Ovoviviparous, producing 4–8 neonates (180–230mm). **VENOM** Mixture of procoagulants; venom anticoagulants and haemorrhagins possibly present. Bites poorly documented but has potential for lethal envenoming. Polyvalent Anti Snake Venom Serum (Central Research Institute, Kasauli); SII Polyvalent Antisnake Venom Serum (lyophilized) (Serum Institute of India Ltd).

Omkoi Lance-headed Pit Viper ■ *Protobothrops kelomohy* 131cm

DESCRIPTION Back of body brown, with 51–53 reddish-brown, black-edged transverse blotches, forming zigzag median line; row of large ventro-lateral blotches on flanks; tail

with 16/15 dark bands; lance-head pattern on forehead; three bold vertical facial stripes; a postocular stripe; belly whitish-brown, with irregular series of brown blotches. **DISTRIBUTION** Sop Khong, Omkoi District, Chiang Mai Province, northern Thailand. **HABITS AND HABITAT** Dry evergreen forest at elevation of 600m asl. Diet and reproductive habits unstudied.

Brown-spotted Pit Viper ■ *Protobothrops mucrosquamatus* 117.4cm

DESCRIPTION Top of body is greyish olive or brown, with a series of large, irregular, dark-edged brown spots; a series of smaller, dark brown spots is present on flanks; a dark post-ocular streak; tail is light brown with black spots; belly is cream with light brown speckles. Body is slender; head is distinct from neck; snout is elongate; forehead scales are reduced; eyes are small with a vertical pupil; dorsals are strongly keeled posteriorly; has long, narrow, non-erect supra-oculars; subcaudals are paired; cloacal scute is entire. **DISTRIBUTION** Myanmar and Thailand. Extralimitally: Laos, Vietnam, Bangladesh, E India, S China. **HABITS AND HABITAT** Temperate and subtropical evergreen forests (250–1,088m asl). Arboreal as well

as terrestrial. Diet comprises frogs, mammals, snakes and birds. Oviparous, laying 5–13 eggs (size unknown). **VENOM** Fibrinogenases and haemorrhagins; venom anticoagulants possibly present. Potential for lethal envenoming. Polyvalent Anti Snake Venom Serum (Central Research Institute, Kasauli); SII Polyvalent Antisnake Venom Serum (lyophilized) (Serum Institute of India Ltd); Bivalent Antivenin Pit Viper, Trimeresurus antivenin (National Institute of Preventative Medicine, Taiwan).

Bornean Palm Pit Viper ■ *Trimeresurus borneensis* 83cm

DESCRIPTION Top of body is variable, ranging from mottled light brown, through mid-

brown with a dark brown saddle-like pattern comprising 20–30 blotches or cross-bars, to bright yellow with darker mottling; an oblique, pale post-ocular stripe extends to neck; belly is paler. Body is robust; head is triangular and distinct from neck; snout is projecting; forehead scales are smooth; eyes are small with a vertical pupil; dorsals are smooth or weakly keeled; tail is prehensile; cloacal scute is entire. **DISTRIBUTION** Borneo. **HABITS AND HABITAT** Lowland swamps and mid-hill forests (<c. 1,130m asl). Terrestrial as juveniles, semi-arboreal as adults. Diet comprises mammals. Oviparous, producing 7–14 eggs (size unknown). **VENOM** Procoagulants, haemorrhagins and necrotoxins possibly present. Bites poorly documented but has potential for serious envenoming. Green Pit Viper Antivenin (Thai Red Cross Society).

Javanese Palm Pit Viper ▪ *Trimeresurus puniceus* 92cm
(Bahasa Indonesia: Ular Gibuk, Ular Badotan Puspo)

DESCRIPTION Top of body is grey or yellowish brown, with 20–30 dark cross-bands; males are dark, with irregular, constricted dorso-lateral blotches and cream and dark dots; in females, the pattern shows less contrast; belly is similar. Body is slender in males, relatively stout in females; head is distinct from neck; snout is distinctly projected and raised, and strongly obliquely truncated; eyes are small with a vertical pupil; dorsals are smooth or weakly keeled; cloacal scute is entire. **DISTRIBUTION** Sumatra, Mentawai Archipelago, Java, Pulau Tinjil. **HABITS AND HABITAT** Lowland and montane forests (<1,600m asl); also tea and coffee plantations. Males are terrestrial, females semi-arboreal. Diet includes small vertebrates. Ovoviviparous, producing 7–33 neonates (180mm). **VENOM** Venom characteristics and bite remain unstudied. Green Pit Viper Antivenin (Thai Red Cross Society).

Wirot's Palm Pit Viper ▪ *Trimeresurus wiroti* 88.9cm
(Thai: Ngu Palm)

DESCRIPTION Top of body is dark greyish brown in males and dark brown in females, with 22–35 darker cross-bands; dark brown dorso-lateral blotches are present, the areas between them darker than on flanks and suffused with dark and light dots or blotches; belly is darker. Body is slender in males, stout in females; head is distinct from neck; has a distinctly projected and raised snout, strongly obliquely truncated; has narrow supra-oculars, convex or granular and raised; eyes are small with a vertical pupil; occipitals, temporals and dorsals are moderately keeled or smooth; subcaudals are paired; cloacal scute is entire. **DISTRIBUTION** Thailand, Peninsular Malaysia. **HABITS AND HABITAT** Lowland and hill forests. Arboreal, on trees. Diet comprises mammals, birds and frogs. Oviparous, producing 7–12 eggs (size unknown). **VENOM** Venom characteristics and bite remain unstudied. No antivenom raised.

Bornean Keeled Green Pit Viper ■ *Tropidolaemus subannulatus* 96.3cm
(Bahasa Malaysia/Indonesia: Ular Kapak. Iban: Ular Engkaradau)

DESCRIPTION Top of body is green or greenish blue, with dark cross-bars that are white or red in males and juveniles, and pale blue in adult females; post-ocular stripe is cream or yellow in adult females, and white and red in juveniles and males; belly is uniform pale green, or blotched or spotted with blue or red. Body is slender in juveniles, relatively thick in adults; head is distinct from neck; eyes are small with a vertical pupil; tail is prehensile; occipitals are distinctly keeled in males; subcaudals are paired; cloacal scute is divided. **DISTRIBUTION** Borneo, Pulau Belitung. Extralimitally: Buton, Sangihe Archipelago, Sulawesi, Balabac, Basilan, Bohol, Dinagat, Jolo, Leyte, Luzon, Mindanao, Negros, Palawan, Panay, Samar, Sibutu, Tumindao. **HABITS AND HABITAT** Lowland forests. Low vegetation and mid-level of trees in riparian forests. Diet comprises birds and rodents. Ovoviviparous (numbers and size of neonates unknown). **VENOM** Mixture of procoagulants; anticoagulants, haemorrhagins and necrotoxins possibly present. Envenoming may be lethal. Green Pit Viper Antivenin (Thai Red Cross Society).

Wagler's Keeled Green Pit Viper ■ *Tropidolaemus wagleri* 92cm
(Bahasa Malaysia: Ular Kapak Tokong. Thai: Ngu Kieo Tuk-kae)

DESCRIPTION In adult and juvenile males, top of body is black with white spots; post-ocular stripe is white and red; belly is pale. In adult females, top of body is black with yellow cross-bars; post-ocular stripe is black; belly is banded. In juvenile

females, top of body is yellow with white cross-bars. Body is slender and laterally compressed in juveniles and males, and robust in adult females; head is triangular and distinct from neck; forehead scales are small and distinctly keeled; internasals are in contact; eyes are small with a vertical pupil; scales on throat are keeled; tail is short and prehensile; dorsals are feebly keeled in males, distinctly keeled in females; subcaudals are paired; cloacal scute is divided. **DISTRIBUTION** Thailand, Peninsular Malaysia, Singapore, Sumatra, Pulau Nias, Mentawai Archipelago, Pulau Bangka, Pulau Natuna, Riau Archipelago. Extralimitally: Vietnam. **HABITS AND HABITAT** Lowland forests. Arboreal, on low vegetation and mid-level of trees. Diet comprises birds and rodents. Ovoviviparous, producing 15–41 neonates (150mm). **VENOM** Mixture of procoagulants; anticoagulants, haemorrhagins and necrotoxins possibly present. Envenoming may be lethal. Green Pit Viper Antivenin (Thai Red Cross Society).

Gumprecht's Pit Viper ▪ *Viridovipera gumprechti* 128cm

DESCRIPTION Top of body is bright pale green in males, dark green in females; has 3 small white vertebral spots at posterior of body; interstitial skin is black; post-ocular streak is red dorsally and white ventrally in males, and white in females; ventro-lateral stripe is white dorsally and red ventrally in males, and white or blue in females; tail is rusty or reddish brown; chin and gular region are emerald-green; belly is emerald-green anteriorly, turning dark bluish green and then reverting to emerald-green posteriorly; iris is red in males, yellow in females. Body is slender, elongate and triangular in cross section; head is distinct from neck; eyes are small with a vertical pupil; tail is long and prehensile; dorsals are moderately keeled; subcaudals are paired;

cloacal scute is entire. **DISTRIBUTION** Myanmar, Thailand. Extralimitally: Laos, Vietnam, E India, S China. **HABITS AND HABITAT** Seasonal and dry subtropical forests, near bamboo thickets and hill streams (800–1,200m asl). Arboreal, in trees; also terrestrial. Diet comprises skinks and rodents. Ovoviviparous, producing 9–15 neonates (size unknown). **VENOM** Venom characteristics and bite remain unstudied. Green Pit Viper Antivenin (Thai Red Cross Society).

Medo Pit Viper ▪ *Viridovipera medoensis* 67.7cm

DESCRIPTION Top of body is dark green, sometimes edged with turquoise-blue; has a white and red ventro-lateral stripe; belly is lighter green or yellowish cream; iris is green or yellowish green. Body is slender, cylindrical and slightly compressed laterally; head is broad and distinct from neck; scales on upper snout are enlarged; eyes are small with a vertical pupil; tail is short and prehensile; dorsals have obtuse keels on vertebral region and flanks; subcaudals are mostly paired; cloacal scute is entire. **DISTRIBUTION** Myanmar. Extralimitally: E India, S China. **HABITS AND HABITAT** Steep slopes of tropical and

subtropical montane forests (1,000–1,400m asl). Arboreal, in bamboo thickets. Diet comprises frogs and rodents. Reproductive biology is unstudied. **VENOM** Venom characteristics and bite remain unstudied. Polyvalent Anti Snake Venom Serum (Central Research Institute, Kasauli); SII Polyvalent Antisnake Venom Serum (lyophilized) (Serum Institute of India Ltd).

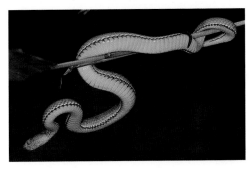

Stejneger's Pit Viper ▪ *Viridovipera stejnegeri* 112cm
(Thai: Ngu Kieo Pi)

DESCRIPTION Top of body and head are bright green; interstitial skin is dark grey or greyish brown; belly is pale or greenish cream; males have a white post-ocular stripe; ventro-lateral stripe is orange, brown or red ventrally and white dorsally in males, and white in females; belly is pale green; iris is red or amber in males, and yellow or amber in females. Body is slender and triangular in cross section; head is distinct from neck; eyes are small with a vertical pupil; tail is long and prehensile; dorsals are weakly keeled; subcaudals are paired; caudal scale is entire. **DISTRIBUTION** Myanmar. Extralimitally: Laos, Vietnam, E

India, China, Taiwan. **HABITS AND HABITAT** Hill forests, near fast-flowing streams (500–2,845m asl). Arboreal, in bushes and low trees. Diet comprises rodents, birds, lizards and frogs. Ovoviviparous, producing 3–10 neonates (155–175mm). **VENOM** Fibrinogenases present, anticoagulants and haemorrhagins possibly present. Severe, potentially lethal envenoming possible. Polyvalent Anti Snake Venom Serum (Central Research Institute, Kasauli); Green Pit Viper Antivenin (Thai Red Cross Society); Bivalent Antivenin Pit Viper, Trimeresurus antivenin (National Institute of Preventative Medicine, Taiwan).

Vogel's Pit Viper ▪ *Viridovipera vogeli* 112cm

DESCRIPTION Top of body is dark green or grass-green (ventro-lateral line is mostly white in adult males of Thai population); interstitial skin is bright blue; males have white vertebral flecks, a yellow or pale orange iris and, sometimes, a post-ocular stripe; females have a yellowish-green ventro-lateral stripe and a tan-yellow or yellowish-green iris; belly is yellowish green; tail tip is green, dark grey or brown. Body is slender and elongate; head is large and distinct from neck; eyes are small with a vertical pupil; dorsals are smooth or

strongly keeled; temporal and rear head scales are weakly keeled or smooth; caudal scale is entire. **DISTRIBUTION** Thailand. Extralimitally: Laos, Cambodia, Vietnam. **HABITS AND HABITAT** Evergreen, semi-evergreen and dry evergreen forests, grasslands and plantations, at elevations (200–1,200m asl). Edges of forest streams. Diet includes small mammals, skinks and insects. Reproductive habits are unstudied. **VENOM** Procoagulants, anticoagulants and haemorrhagins possibly present. Severe, potentially lethal envenoming possible. Green Pit Viper Antivenin (Thai Red Cross Society).

> **XENODERMATIDAE – STRANGE-SKINNED SNAKES**
> The members of this family arguably include some of the least-known species in the region. They have scales that are nearly completely fused to the underlying skin, and live in lowland forests of Southeast Asia.

Stoliczka's Stream Snake ■ *Stoliczkia borneensis* 75cm

DESCRIPTION Top of body is bluish brown or mid-brown, with dark squarish marks that are as broad as or broader than their interspaces; several dark bars are present behind forehead; belly is plain brown. Body is slender and laterally compressed, with a sharp ridge on vertebral region; head is large and distinct from neck; forehead has 2 rows of small scales in front of eyes; nostrils are large and flaring; eyes are small and beady with a vertical pupil; tail is long and slender; dorsals are strongly keeled; vertebrals are enlarged; subcaudals are single; caudal scale is entire. **DISTRIBUTION** Borneo. **HABITS AND HABITAT** Sub-montane and montane forests (800–1,800m). On stream banks, as well as low vegetation. Diet and reproductive habits are unstudied.

Rough-backed Litter Snake ■ *Xenodermus javanicus* 65cm
(Thai: Ngu Tong Kao)

DESCRIPTION Top of body is plain grey; ridges on scales are cream; snout is slightly paler than rest of head; labials and throat are cream; belly is cream with black areas. Body is slender and compressed; head is large and distinct from neck; 3 rows of large keeled scales are present on dorsum; nostrils are flaring and pointed forward; eyes are large with a rounded pupil; tail is long; subcaudals are single; caudal scale is entire. **DISTRIBUTION** Myanmar, Thailand, Peninsular Malaysia, Sumatra, Borneo, Java. **HABITS AND HABITAT** Lowland dipterocarp forests and agricultural areas (500–1,100m asl). Terrestrial and sub-fossorial. Diet comprises frogs. Oviparous, clutches comprising 2–4 eggs (23–28 × 9–11mm).

XENOPHIDIIDAE SPINE – JAWED SNAKES
The two known members of this family are characterized by compressed bodies, short tails, rounded pupils, enlarged prefrontals, the absence of a loreal scute, undivided nasals and a single anal scute. The right lung is not vascularized, a pelvis is absent and there is an ectopterygoid process on the maxilla.

Bornean Spiny-jawed Snake ■ *Xenophidion acanthognathus* 335mm

DESCRIPTION Top of body brown with zigzag longitudinal pattern; forehead mid-brown with pale areas; broad white postocular stripe extends up to sides of neck; zigzag tan stripes on flanks; belly black with small, squarish yellow spots. Body slender and strongly compressed; head slightly flattened and indistinct from neck; loreal absent; scales keeled (except two lowest rows on body). **DISTRIBUTION** Borneo. **HABITAT AND HABITS** Found in forested mid-hills at about 600m above sea level. Possibly subfossorial. Diet includes skinks. Oviparous; other details of reproductive habits unknown.

Typhlopidae – Blind Snakes
Blind snakes are readily distinguished from all other snakes, although some members are occasionally confused with earthworms. They show blunt heads, short tails, small eyes that are concealed under scales, ventrals that are not enlarged, and a toothless lower jaw. They tend to be nocturnal and sub-fossorial species, feeding on small arthropods – mainly termites and ants, and their eggs, larvae and pupae. They are mainly oviparous, and have a cosmopolitan distribution in tropical and subtropical regions.

Diard's Blind Snake ▪ *Argyrophis diardii* 43cm

DESCRIPTION Top of body is dark brown, each scale with an indistinct, light transverse streak; belly and sides are light brown, with a gradual transition to dark dorsum. Body is moderately thick; head is indistinct from neck; snout is rounded and strongly projecting; upper rostral is narrow, widening ventrally; eyes are distinct; scales are smooth; total length is 30–50 times mid-body diameter; a caudal spine is present. **DISTRIBUTION** Myanmar, Thailand. Extralimitally: Pakistan, Bangladesh, India, Nepal, Cambodia, Laos, Vietnam, S China. **HABITS AND HABITAT** Mid-hill forests. Sub-fossorial. Diet comprises insects and their larvae. Oviparous, laying 4–17 eggs (4–7mm long).

Müller's Blind Snake ■ *Argyrophis muelleri* 54cm
(Thai: Ngu-din Yai Malayu)

DESCRIPTION Top of body is dark brown, purple or black; belly is yellow or gold, clearly demarcated from top colour. Body is moderately built; head is indistinct from neck; snout is rounded and strongly projecting; upper rostral is narrow, widening ventrally; eyes are

distinct; scales are smooth; total length is 25–45 times mid-body diameter; a caudal spine is present. **DISTRIBUTION** Myanmar, Thailand, Peninsular Malaysia, Singapore, Sumatra, Pulau Bangka, Pulau Nias, Pulau Weh, Pulau Nias, Borneo. Extralimitally: Laos, Cambodia, Vietnam, New Guinea. **HABITS AND HABITAT** Forested lowlands and mid-hills, and areas of wet agriculture (<1,676m). Sub-fossorial, in waterlogged areas, under rocks and logs. Diet includes larvae of ants and termites. Ovoviviparous, producing 2–22 neonates (size unknown).

Thai Blind Snake ■ *Argyrophis siamensis* 17cm

DESCRIPTION Body greyish-olive, scales with a pale yellow-brown anterior edge; position of eyes indicated by a darkened area; belly yellow. **DISTRIBUTION** Thailand. Extralimitally: Cambodia and Vietnam. **HABITS AND HABITAT** Inhabits lowland forests, in leaf litter. Diet presumably comprises small arthropods and worms.

Brahminy Blind Snake ■ *Indotyphlops braminus* 18cm
(Bahasa Malaysia: Ular Buta Biasa. Bahasa Indonesia: Ular Buta Brahminy.
Thai: Ngu-din Ban)

DESCRIPTION Top of body is black or brown; belly is lighter; snout and tip of tail are paler than rest of dorsum. Body is moderately built; head is indistinct from neck; snout is rounded; nostrils are situated laterally; eyes are distinct; total length is 30–45 times mid-body diameter; a caudal spine is present. **DISTRIBUTION** Myanmar, Thailand, Peninsular Malaysia, Singapore, Sumatra, Borneo, Java, Bali. Extralimitally: widespread in tropical, subtropical and temperate regions worldwide. **HABITS AND HABITAT** Lightly forested areas; also human habitations (<2,000m asl). Sub-fossorial. Diet includes termites and ants, and their larvae. All-female species; parthenogenetic, producing 1–8 eggs (11–20mm).

Jerdon's Blind Snake ■ *Indotyphlops jerdoni* 28cm

DESCRIPTION Top of body is dark brown to nearly black; belly is light brown; snout and anal region are cream. Body is moderately built; head is indistinct from neck; snout is rounded; rostral is less than half width of head; eyes are distinct; scales are smooth; total length is 35–50 times mid-body diameter; tail is bluntly pointed; a caudal spine is present. **DISTRIBUTION** Myanmar. Extralimitally: Bhutan, Nepal, E India. **HABITS AND HABITAT** Evergreen forests in mid-hills (<855m asl). Sub-fossorial, under boulders and inside dead trees. Diet is unstudied. Probably oviparous, with a clutch size of 8 (egg size unknown).

Olive Blind Snake ■ *Ramphotyphlops olivaceus* 410mm

DESCRIPTION Body slender; head indistinct from neck; snout projecting, with narrow transverse edge; rostral scute large, over half head width; eyes distinct; caudal spine present. Top of body pale brown; belly paler. **DISTRIBUTION** Borneo. Extralimitally: Sangihe Archipelago, Seram and Mysool (Indonesia), and Samar and Babuan (the Philippines). **HABITAT AND HABITS** Found in lowland rainforests. Fossorial, living in leaf litter. Diet and reproductive habits unknown.

Global Status According to the IUCN Red List 2020 (version 2020-2)

Extinct (EX) 'A taxon is Extinct when there is no reasonable doubt that the last individual has died. A taxon is presumed Extinct when exhaustive surveys in known and/or expected habitat, at appropriate times (diurnal, seasonal, annual), throughout its historic range have failed to record an individual. Surveys should be over a time frame appropriate to the taxon's life cycle and life form.'

Extinct in the Wild (EW) 'A taxon is Extinct in the Wild when it is known only to survive in cultivation, in captivity or as a naturalized population (or populations) well outside the past range. A taxon is presumed Extinct in the Wild when exhaustive surveys in known and/or expected habitat, at appropriate times (diurnal, seasonal, annual), throughout its historic range have failed to record an individual. Surveys should be over a time frame appropriate to the taxon's life cycle and life form.'

Critically Endangered (CR) 'A taxon is Critically Endangered when the best available evidence indicates that it meets any of the criteria A to E for Critically Endangered, and it is therefore considered to be facing an extremely high risk of extinction in the wild.'

Endangered (EN) 'A taxon is Endangered when the best available evidence indicates that it meets any of the criteria A to E for Endangered, and it is therefore considered to be facing a very high risk of extinction in the wild.'

Vulnerable (VU) 'A taxon is Vulnerable when the best available evidence indicates that it meets any of the criteria A to E for Vulnerable, and it is therefore considered to be facing a high risk of extinction in the wild.'

Near Threatened (NT) 'A taxon is Near Threatened when it has been evaluated against the criteria but does not qualify for Critically Endangered, Endangered or Vulnerable now, but is close to qualifying for or is likely to qualify for a threatened category in the near future.'

Least Concern (LC) 'A taxon is Least Concern when it has been evaluated against the criteria and does not qualify for Critically Endangered, Endangered, Vulnerable or Near Threatened. Widespread and abundant taxa are included in this category.'

Data Deficient (DD) 'A taxon is Data Deficient when there is inadequate information to make a direct, or indirect, assessment of its risk of extinction based on its range and/or population status. A taxon in this category may be well studied, and its biology well known, but appropriate data on abundance and/or range are lacking. Data Deficient is therefore not a category of threat. Listing of taxa in this category indicates that more information is required and acknowledges the possibility that future research will show that threatened classification is appropriate.'

Not Evaluated (NE) 'A taxon is Not Evaluated when it is has not yet been evaluated against the criteria.'

Symbols
- E   Endemic to the region and/or associated islands
- +   Present
- –   Absent
- ?   Suspected to occur

Common English Name	Scientific Name	Myanmar	Thailand	Peninsular Malaysia	Sumatra	Borneo	Java	Bali	Global Status
Acrochordidae (Wart Snakes)									
Wart Snake	Acrochordus granulatus	+	+	+	+	+	+	–	LC
Elephant Trunk Snake	Acrochordus javanicus	–	+	+	+	+	+	–	LC
Anomochilidae (Giant Blind Snakes)									
Malayan Giant Blind Snake	Anomochilus leonardi	–	–	+	–	+	–	–	LC
Kinabalu Giant Blind Snake	Anomochilus monticola	–	–	–	–	E	–	–	DD
Sumatran Giant Blind Snake	Anomochilus weberi	–	–	–	+	+	–	–	DD
Cylindrophiidae (Pipe Snakes)									
Burmese Pipe Snake	Cylindrophis burmanus	+	–	–	–	–	–	–	NE
Engkari Pipe Snake	Cylindrophis engkariensis	–	–	–	–	E	–	–	DD
Lined Pipe Snake	Cylindrophis lineatus	–	–	–	–	E	–	–	DD
Common Pipe Snake	Cylindrophis ruffus	+	+	+	+	+	+	–	LC
Javanese Pipe Snake	Cylindrophis subocularis	–	–	–	–	–	+	–	NE
Pythonidae (Pythons)									
Reticulated Python	Malayopython reticulatus	+	+	+	+	+	+	+	NE
Southeast Asian Rock Python	Python bivittatus	+	+	?	–	–	+	–	V
Bornean Short Python	Python breitensteini	–	–	–	–	E	–	–	LC
Brongersma's Short Python	Python brongersmai	–	+	+	+	–	–	–	LC
Sumatran Short Python	Python curtus	–	–	–	E	–	–	–	NE
Burmese Short Python	Python kyaiktiyo	E	–	–	–	–	–	–	V
Xenopeltidae (Sunbeam Snakes)									
Sunbeam Snake	Xenopeltis unicolor	+	+	+	+	+	+	–	LC
Colubridae ('Typical' Snakes)									
Speckle-headed Vine Snake	Ahaetulla fasciolata	–	+	+	+	+	–	–	LC
River Vine Snake	Ahaetulla fronticincta	E	–	–	–	–	–	–	LC
Malayan Vine Snake	Ahaetulla mycterizans	–	+	+	+	–	+	–	LC
Long-nosed Vine Snake	Ahaetulla nasuta	+	+	–	–	–	–	–	NE
Oriental Vine Snake	Ahaetulla prasina	+	+	+	+	+	+	+	LC
Dice-like Rat Snake	Archelaphe bella	+	–	–	–	–	–	–	LC
Iridescent Snake	Blythia reticulata	+	–	–	–	–	–	–	NE

▪ CHECKLIST OF SOUTHEAST ASIAN SNAKES ▪

Common English Name	Scientific Name	Myanmar	Thailand	Peninsular Malaysia	Sumatra	Borneo	Java	Bali	Global Status
Bengkulu Cat Snake	Boiga bengkuluensis	–	+	–	+	–	–	–	DD
Green Cat Snake	Boiga cyanea	+	+	+	–	–	–	–	NE
Dog-toothed Cat Snake	Boiga cynodon	–	+	+	+	+	+	+	LC
Mangrove Cat Snake	Boiga dendrophila	–	+	+	+	+	+	+	NE
White-spotted Cat Snake	Boiga drapiezii	–	+	+	+	+	+	–	LC
Jasper Cat Snake	Boiga jaspidea	–	+	+	+	+	+	–	NE
Many-spotted Cat Snake	Boiga multomaculata	+	+	–	–	–	–	–	NE
Black-headed Cat Snake	Boiga nigriceps	–	+	+	+	+	+	–	LC
Tawny Cat Snake	Boiga ochracea	+	+	–	–	–	–	–	NE
Assamese Cat Snake	Boiga quincunciata	+	–	–	–	–	–	–	NE
Banded Green Cat Snake	Boiga saengsomi	–	E	–	–	–	–	–	EN
Thai Cat Snake	Boiga siamensis	+	+	–	–	–	–	–	NE
Wall's Cat Snake	Boiga walli	+	–	–	–	–	–	–	LC
Padang Reed Snake	Calamaria abstrusa	–	–	–	E	–	–	–	DD
White-bellied Reed Snake	Calamaria albiventer	–	–	+	+	–	–	–	LC
Bengkulu Reed Snake	Calamaria alidae	–	–	–	E	–	–	–	DD
Battersby's Reed Snake	Calamaria battersbyi	–	–	–	–	+	–	–	DD
Bicoloured Reed Snake	Calamaria bicolor	–	–	–	–	+	+	–	LC
Bornean Reed Snake	Calamaria borneensis	–	–	–	–	+	–	–	LC
Thick Reed Snake	Calamaria crassa	–	–	–	E	–	–	–	DD
Döderlein's Reed Snake	Calamaria doederleini	–	–	–	E	–	–	–	DD
Eiselt's Reed Snake	Calamaria eiselti	–	–	–	E	–	–	–	DD
Everett's Reed Snake	Calamaria everetti	–	–	–	–	E	–	–	LC
Forcart's Reed Snake	Calamaria forcarti	–	–	–	E	–	–	–	DD
Gimlett's Reed Snake	Calamaria gimletti	–	–	+	–	–	–	–	LC
Grabowsky's Reed Snake	Calamaria grabowskyi	–	–	–	–	E	–	–	LC
Slender Reed Snake	Calamaria gracillima	–	–	–	–	E	–	–	DD
Lined Reed Snake	Calamaria griswoldi	–	–	–	–	E	–	–	LC
Hillenius's Reed Snake	Calamaria hilleniusi	–	–	–	–	E	–	–	LC
Inger's Reed Snake	Calamaria ingeri	–	–	E	–	–	–	–	CR
Javanese Reed Snake	Calamaria javanica	–	–	–	–	–	+	–	DD
White-striped Reed Snake	Calamaria lateralis	–	–	–	–	+	+	–	DD
Collared Reed Snake	Calamaria leucogaster	–	–	–	+	+	–	–	LC
Linnaeus's Reed Snake	Calamaria linnaei	–	–	–	–	–	+	–	LC
Low's Reed Snake	Calamaria lovii	–	+	+	+	–	+	–	LC
Variable Reed Snake	Calamaria lumbricoidea	–	+	+	+	+	+	–	LC
Lumholtz's Reed Snake	Calamaria lumholtzi	–	–	–	–	E	–	–	DD
Stripe-necked Reed Snake	Calamaria margaritophora	–	–	–	E	–	–	–	DD
Mechel's Reed Snake	Calamaria mecheli	–	–	–	E	–	–	–	DD

▪ Checklist of Southeast Asian Snakes ▪

Common English Name	Scientific Name	Myanmar	Thailand	Peninsular Malaysia	Sumatra	Borneo	Java	Bali	Global Status
Kapuas Reed Snake	Calamaria melanota	–	–	–	–	E	–	–	LC
Yellow-spotted Reed Snake	Calamaria modesta	–	–	–	+	+	+	–	LC
Brown Reed Snake	Calamaria pavimentata	+	+	–	–	–	–	–	LC
Prakke's Reed Snake	Calamaria prakkei	–	–	+	–	+	–	–	CR
Rebentisch's Reed Snake	Calamaria rebentischi	–	–	–	–	E	–	–	DD
Red-headed Reed Snake	Calamaria schlegeli	–	+	+	+	+	+	+	LC
Schmidt's Reed Snake	Calamaria schmidti	–	–	–	–	E	–	–	LC
Yellow-bellied Reed Snake	Calamaria suluensis	–	–	–	–	+	–	–	LC
Sumatran Reed Snake	Calamaria sumatrana	–	–	–	E	–	–	–	LC
Ulmer's Reed Snake	Calamaria ulmeri	–	–	–	E	–	–	–	DD
Short-tailed Reed Snake	Calamaria virgulata	–	–	–	+	+	+	–	LC
Ornate Flying Snake	Chrysopelea ornata	+	+	+	–	–	–	–	NE
Garden Flying Snake	Chrysopelea paradisi	+	+	+	+	+	+	–	NE
Twin-barred Flying Snake	Chrysopelea pelias	+	+	+	+	+	+	–	LC
Enggano Rat Snake	Coelognathus enganensis	–	–	–	E	–	–	–	DD
Philippine Rat Snake	Coelognathus erythrurus	–	–	–	–	+	–	–	NE
Yellow-striped Rat Snake	Coelognathus flavolineatus	+	+	+	+	+	+	–	LC
Copper-head Rat Snake	Coelognathus radiatus	+	+	+	+	+	+	–	NE
Indonesian Rat Snake	Coelognathus subradiatus	–	–	–	–	–	–	?	NE
Mountain Dwarf Snake	Collorhabdium williamsoni	–	–	E	–	–	–	–	LC
Doria's Green Snake	Cyclophiops doriae	+	–	–	–	–	–	–	NE
Hampton's Green Snake	Cyclophiops hamptoni	E	–	–	–	–	–	–	DD
Twin-lored Bronzeback Tree Snake	Dendrelaphis biloreatus	+	–	–	–	–	–	–	LC
Stripe-tailed Bronzeback Tree Snake	Dendrelaphis caudolineatus	+	+	+	+	+	–	–	NE
Blue Bronzeback Tree Snake	Dendrelaphis cyanochloris	+	+	+	–	–	–	–	LC
Beautiful Bronzeback Tree Snake	Dendrelaphis formosus	–	+	+	+	+	+	–	LC
Haas's Bronzeback Tree Snake	Dendrelaphis haasi	–	–	+	+	+	+	–	LC
Kopstein's Bronzeback Tree Snake	Dendrelaphis kopsteini	–	+	+	+	+	–	–	LC
Saw-tooth-necked Bronzeback Tree Snake	Dendrelaphis nigroserratus	+	+	–	–	–	–	–	NE
Painted Bronzeback Tree Snake	Dendrelaphis pictus	–	+	+	+	+	+	+	NE
Northwestern Painted Bronzeback Tree Snake	Dendrelaphis proarchos	+	–	–	–	–	–	–	NE
Striated Bronzeback Tree Snake	Dendrelaphis striatus	–	+	+	+	+	–	–	LC
Mountain Bronzeback Tree Snake	Dendrelaphis subocularis	+	+	–	–	–	–	–	LC
Indian Bronzeback Tree Snake	Dendrelaphis tristis	+	–	–	–	–	–	–	NE
Underwood's Bronzeback Tree Snake	Dendrelaphis underwoodi	–	–	–	–	–	E	–	DD
Wall's Bronzeback Tree Snake	Dendrelaphis walli	E	–	–	–	–	–	–	LC
Davison's Bridled Snake	Dryocalamus davisonii	+	+	–	–	–	–	–	LC

Common English Name	Scientific Name	Myanmar	Thailand	Peninsular Malaysia	Sumatra	Borneo	Java	Bali	Global Status
Half-banded Bridled Snake	Dryocalamus subannulatus	–	+	+	+	+	–	–	LC
Three-banded Bridled Snake	Dryocalamus tristrigatus	–	–	–	–	+	–	–	LC
Keel-bellied Whip Snake	Dryophiops rubescens	–	+	+	+	+	+	–	LC
Eastern Rat Snake	Elaphe cantoris	+	–	–	–	–	–	–	NE
Cave Racer	Elaphe taeniurus	+	+	+	+	+	–	–	NE
Dark Grey Ground Snake	Elapoidis fusca	–	–	–	+	+	+	–	LC
Sumatran Burrowing Snake	Etheridgeum pulchrum	–	–	–	E	–	–	–	DD
Mandarin Rat Snake	Euprepiophis mandarinus	+	–	–	–	–	–	–	NE
Orange-bellied Snake	Gongylosoma baliodeirum	–	+	+	+	+	+	–	LC
Striped Ground Snake	Gongylosoma longicauda	–	–	+	+	+	+	–	LC
Pulau Tioman Ground Snake	Gongylosoma mukutense	–	–	E	–	–	–	–	CR
Indo-Chinese Ground Snake	Gongylosoma scriptum	+	+	–	–	–	–	–	LC
Khasi Hills Trinket Snake	Gonyosoma frenatum	+	–	–	–	–	–	–	NE
Royal Tree Snake	Gonyophis margaritatus	–	–	+	–	+	–	–	LC
Green Trinket Snake	Gonyosoma prasinum	+	+	+	–	–	–	–	LC
Red-tailed Racer	Gonyosoma oxycephalum	+	+	+	+	+	+	–	LC
Stripe-necked Snake	Liopeltis frenatus	+	–	–	–	–	–	–	LC
Stoliczka's Ringneck	Liopeltis stoliczkae	+	–	–	–	–	–	–	LC
Tricoloured Ringneck	Liopeltis tricolor	–	+	+	+	+	+	–	LC
Dusky Wolf Snake	Lycodon albofuscus	–	–	+	+	+	–	–	NE
Indian Wolf Snake	Lycodon aulicus	+	–	–	–	–	–	–	NE
Butler's Wolf Snake	Lycodon butleri	–	–	+	–	–	–	–	LC
Island Wolf Snake	Lycodon capucinus	+	+	+	+	+	+	+	LC
Cardamom Mountains Wolf Snake	Lycodon cardamomensis	–	+	–	–	–	–	–	DD
Gua Wang Burma Wolf Snake	Lycodon cavernicolus	–	–	+	–	–	–	–	NE
Brown Wolf Snake	Lycodon effraenis	–	–	+	+	+	–	–	LC
Banded Wolf Snake	Lycodon fasciatus	+	+	–	–	–	–	–	NE
Yellow Wolf Snake	Lycodon flavozonatus	+	–	–	–	–	–	–	LC
Yellow-speckled Wolf Snake	Lycodon jara	+	–	–	–	–	–	–	LC
Kundu's Wolf Snake	Lycodon kundui	E	–	–	–	–	–	–	DD
Laos Wolf Snake	Lycodon laoensis	–	+	+	–	–	–	–	LC
Snake-eating Wolf Snake	Lycodon ophiophagus	–	E	–	–	–	–	–	LC
Northern Wolf Snake	Lycodon septentrionalis	+	+	–	–	–	–	–	NE
Sidik's Wolf Snake	Lycodon sidiki	–	–	–	+	–	–	–	NE
White-banded Wolf Snake	Lycodon subcinctus	+	+	+	+	+	+	–	LC
Zaw's Wolf Snake	Lycodon zawi	+	–	–	–	–	–	–	LC
Chan-ard's Mountain Reed Snake	Macrocalamus chanardi	–	–	E	–	–	–	–	LC
Golden-bellied Reed Snake	Macrocalamus emas	–	E	–	–	–	–	–	NE

Common English Name	Scientific Name	Myanmar	Thailand	Peninsular Malaysia	Sumatra	Borneo	Java	Bali	Global Status
Genting Highlands Reed Snake	*Macrocalamus gentingensis*	–	–	E	–	–	–	–	LC
Jason's Reed Snake	*Macrocalamus jasoni*	–	–	E	–	–	–	–	LC
Striped Reed Snake	*Macrocalamus lateralis*	–	+	+	–	–	–	–	LC
Schulz's Reed Snake	*Macrocalamus schulzi*	–	–	E	–	–	–	–	LC
Tweedie's Reed Snake	*Macrocalamus tweediei*	–	–	E	–	–	–	–	LC
Vogel's Reed Snake	*Macrocalamus vogeli*	–	–	E	–	–	–	–	LC
White-barred Kukri Snake	*Oligodon albocinctus*	+	–	–	–	–	–	–	NE
Spotted Kukri Snake	*Oligodon annulifer*	–	–	–	–	E	–	–	LC
Barron's Kukri Snake	*Oligodon barroni*	–	+	–	–	–	–	–	LC
Javanese Kukri Snake	*Oligodon bitorquatus*	–	–	–	–	–	+	+	LC
Boo Liat's Kukri Snake	*Oligodon booliati*	–	–	E	–	–	–	–	CR
Chain-banded Kukri Snake	*Oligodon catenatus*	+	+	–	–	–	–	–	NE
Grey Kukri Snake	*Oligodon cinereus*	+	+	–	–	–	–	–	LC
Pegu Kukri Snake	*Oligodon cruentatus*	E	–	–	–	–	–	–	LC
Cantor's Kukri Snake	*Oligodon cyclurus*	+	–	–	–	–	–	–	LC
Deuve's Kukri Snake	*Oligodon deuvei*	–	+	–	–	–	–	–	NE
Spot-tailed Kukri Snake	*Oligodon dorsalis*	+	–	–	–	–	–	–	NE
Jewelled Kukri Snake	*Oligodon everetti*	–	–	–	–	E	–	–	LC
Small-banded Kukri Snake	*Oligodon fasciolatus*	+	+	–	–	–	–	–	NE
Hampton's Kukri Snake	*Oligodon hamptoni*	E	–	–	–	–	–	–	DD
Hua Hin Kukri Snake	*Oligodon huahin*	–	+	–	–	–	–	–	NE
Unicoloured Kukri Snake	*Oligodon inornatus*	–	+	–	–	–	–	–	LC
Jintakune's Kukri Snake	*Oligodon jintakunei*	–	E	–	–	–	–	–	DD
Joynson's Kukri Snake	*Oligodon joynsoni*	–	+	–	–	–	–	–	LC
Arakan Kukri Snake	*Oligodon mcdougalli*	E	–	–	–	–	–	–	LC
Meyerink's Kukri Snake	*Oligodon meyerinkii*	–	–	–	–	+	–	–	EN
Cambodian Kukri Snake	*Oligodon mouhoti*	–	+	–	–	–	–	–	LC
Eight-lined Kukri Snake	*Oligodon octolineatus*	–	–	+	+	+	+	–	LC
Petronella's Kukri Snake	*Oligodon petronellae*	–	–	–	E	–	–	–	DD
Flat-headed Kukri Snake	*Oligodon planiceps*	E	–	–	–	–	–	–	LC
Javanese Kukri Snake	*Oligodon propinquus*	–	–	–	–	–	E	–	NE
False Striped Kukri Snake	*Oligodon pseudotaeniatus*	–	E	–	–	–	–	–	LC
Padang Kukri Snake	*Oligodon pulcherrimus*	–	–	–	E	–	–	–	VU
Purple Kukri Snake	*Oligodon purpurascens*	–	+	+	+	+	+	–	LC
Four-lined Kukri Snake	*Oligodon quadrilineatus*	+	+	–	–	–	–	–	NE
Sai Yok Kukri Snake	*Oligodon saiyok*	–	+	–	–	–	–	–	NE
Half-keeled Kukri Snake	*Oligodon signatus*	–	–	+	–	+	–	–	LC
Splendid Kukri Snake	*Oligodon splendidus*	E	–	–	–	–	–	–	LC
Striped Kukri Snake	*Oligodon taeniatus*	–	+	–	–	–	–	–	LC

Common English Name	Scientific Name	Myanmar	Thailand	Peninsular Malaysia	Sumatra	Borneo	Java	Bali	Global Status
Mandalay Kukri Snake	Oligodon theobaldi	+	–	–	–	–	–	–	LC
Garlanded Kukri Snake	Oligodon torquatus	E	–	–	–	–	–	–	DD
Three-lined Kukri Snake	Oligodon trilineatus	–	–	–	E	–	–	–	LC
Dark-spined Kukri Snake	Oligodon vertebralis	–	–	–	–	E	–	–	DD
Mountain Reed Snake	Oreocalamus hanitschi	–	–	+	–	+	–	–	LC
Red Bamboo Rat Snake	Oreocryptophis porphyraceus	+	+	+	+	–	–	–	NE
Boonsong's Keelback	Parahelicops boonsongi	–	E	–	–	–	–	–	DD
White-collared Dwarf Reed Snake	Pseudorabdion albonuchalis	–	–	–	–	E	–	–	LC
Mocquard's Dwarf Reed Snake	Pseudorabdion collaris	–	–	–	–	E	–	–	LC
Eiselt's Dwarf Reed Snake	Pseudorabdion eiselti	–	–	–	E	–	–	–	LC
Common Dwarf Reed Snake	Pseudorabdion longiceps	–	+	+	+	+	–	–	LC
Sarawak Reed Snake	Pseudorabdion saravacensis	–	–	–	–	+	–	–	LC
Keeled Rat Snake	Ptyas carinata	+	+	+	+	+	+	–	LC
White-bellied Rat Snake	Ptyas fusca	–	+	+	+	+	–	–	LC
Javanese Rat Snake	Ptyas korros	+	+	+	+	+	+	–	NE
Indian Rat Snake	Ptyas mucosa	+	+	+	+	–	–	–	NE
Green Rat Snake	Ptyas nigromarginata	+	+	–	–	–	–	–	NE
Twin-streaked Black-headed Snake	Sibynophis bistrigatus	E	–	–	–	–	–	–	DD
Collared Black-headed Snake	Sibynophis collaris	+	+	+	–	–	–	–	LC
Striped Black-headed Snake	Sibynophis geminatus	–	+	+	+	+	+	+	LC
White-lipped Black-headed Snake	Sibynophis melanocephalus	–	+	+	+	+	–	–	LC
Triangled Black-headed Snake	Sibynophis triangularis	–	+	–	–	–	–	–	NE
Two-coloured Forest Snake	Smithophis bicolor	+	–	–	–	–	–	–	NE
Bornean Black Snake	Stegonotus borneensis	–	?	–	–	+	–	–	LC
Fruhstorfer's Mountain Snake	Tetralepis fruhstorferi	–	–	–	–	–	E	–	VU
Ornate Brown Snake	Xenelaphis ellipsifer	–	–	+	+	+	–	–	LC
Malayan Brown Snake	Xenelaphis hexagonotus	+	+	+	+	+	+	–	LC
Lamprophiidae (Afro-Eurasian Snakes)									
Indo-Chinese Sand Snake	Psammophis indochinensis	+	+	–	–	–	+	+	LC
Natricidae (Water Snakes)									
Buff-striped Keelback	Amphiesma stolatum	+	+	–	–	–	–	–	NE
Sumatran Keelback	Anoplohydrus aemulans	–	–	–	E	–	–	–	DD
Bornean Keelback	Hebius arquus	–	–	–	–	E	–	–	DD
Two-lined Keelback	Hebius bitaeniatum	+	+	–	–	–	–	–	LC
Boulenger's Keelback	Hebius boulengeri	–	+	–	–	–	–	–	LC
Deschauensee's Keelback	Hebius deschauenseei	–	+	–	–	–	–	–	LC
White-fronted Keelback	Hebius flavifrons	–	–	–	–	E	–	–	LC
Bridled Keelback	Hebius frenatum	–	–	–	–	E	–	–	DD

▪ CHECKLIST OF SOUTHEAST ASIAN SNAKES ▪

Common English Name	Scientific Name	Myanmar	Thailand	Peninsular Malaysia	Sumatra	Borneo	Java	Bali	Global Status
Groundwater's Keelback	Hebius groundwateri	–	E	–	–	–	–	–	DD
Gunung Inas Keelback	Hebius inas	–	+	+	–	–	–	–	LC
Gunung Kerinchi Keelback	Hebius kerinciense	–	–	–	E	–	–	–	DD
Khasi Hills Keelback	Hebius khasiense	+	+	–	–	–	–	–	NE
Günther's Keelback	Hebius modestum	+	+	–	–	–	–	–	LC
Striped Keelback	Hebius parallelum	+	–	–	–	–	–	–	NE
Peters' Keelback	Hebius petersii	–	–	+	+	+	–	–	LC
Red Mountain Keelback	Hebius sanguineum	–	–	E	–	–	–	–	LC
Sarawak Keelback	Hebius saravacense	–	–	+	–	+	–	–	LC
Taron Keelback	Hebius taronense	E	–	–	–	–	–	–	NT
Venning's Keelback	Hebius venningi	+	–	–	–	–	–	–	LC
Viper-like Keelback	Hebius viperinum	–	–	–	E	–	–	–	DD
Strange-tailed Keelback	Hebius xenura	+	–	–	–	–	–	–	NE
Siebold's Keelback	Herpetoreas sieboldii	+	–	–	–	–	–	–	DD
Yellow-spotted Water Snake	Hydrablabes periops	–	–	–	–	E	–	–	LC
Kinabalu Water Snake	Hydrablabes praefrontalis	–	–	–	–	E	–	–	DD
Iguana-jawed Snake	Iguanognathus werneri	–	–	–	E	–	–	–	DD
Spotted Mountain Stream Snake	Opisthotropis maculosa	–	E	–	–	–	–	–	DD
Sumatran Stream Snake	Opisthotropis rugosa	–	–	–	E	–	–	–	LC
Spencer's Stream Snake	Opisthotropis spenceri	–	E	–	–	–	–	–	DD
Corrugated Water Snake	Opisthotropis typica	–	–	–	–	E	–	–	LC
Brown Stream Snake	Paratapinophis praemaxillaris	–	+	–	–	–	–	–	LC
Painted Mock Viper	Psammodynastes pictus	–	–	+	+	+	–	–	NE
Mock Viper	Psammodynastes pulverulentus	+	+	+	+	+	+	–	NE
Günther's Keelback	Rhabdophis chrysargoides	–	–	–	–	–	+	–	DD
Speckle-bellied Keelback	Rhabdophis chrysargos	+	+	+	+	+	+	+	LC
Red-bellied Keelback	Rhabdophis conspicillatus	–	–	+	+	+	–	–	LC
Orange-lipped Keelback	Rhabdophis flaviceps	–	+	+	+	+	–	–	LC
Himalayan Keelback	Rhabdophis himalayanus	+	–	–	–	–	–	–	NE
Leonard's Keelback	Rhabdophis leonardi	+	–	–	–	–	–	–	LC
Gunung Murud Keelback	Rhabdophis murudensis	–	–	–	–	E	–	–	LC
Black-banded Keelback	Rhabdophis nigrocinctus	+	+	–	–	–	–	–	LC
Collared Keelback	Rhabdophis nuchalis	+	–	–	–	–	–	–	LC
Olive Keelback	Rhabdophis plumbicolor	+	–	–	–	–	–	–	LC
Blue-necked Keelback	Rhabdophis rhodomelas	–	+	+	+	+	+	–	LC
Red-necked Keelback	Rhabdophis subminiatus	+	+	+	+	+	+	–	LC
Chinese Keelback Water Snake	Trimerodytes percarinata	+	+	–	–	–	–	–	LC

Common English Name	Scientific Name	Myanmar	Thailand	Peninsular Malaysia	Sumatra	Borneo	Java	Bali	Global Status
Yunnan Keelback Water Snake	Trimerodytes yunnanensis	+	-	-	-	-	-	-	LC
Burmese Keelback Water Snake	Xenochrophis bellula	E	-	-	-	-	-	-	DD
Yellow-spotted Keelback Water Snake	Xenochrophis flavipunctatus	+	+	+	-	-	-	-	LC
Malayan Spotted Keelback Water Snake	Xenochrophis maculatus	-	-	+	+	+	-	-	LC
Javanese Keelback Water Snake	Xenochrophis melanzostus	-	-	-	-	-	+	+	LC
Chequered Keelback Water Snake	Xenochrophis piscator	+	+	-	-	-	-	-	NE
Spotted Keelback Water Snake	Xenochrophis punctulatus	+	+	-	-	-	-	-	LC
St John's Keelback Water Snake	Xenochrophis sanctijohannis	+	-	-	-	-	-	-	NE
Red-sided Keelback Water Snake	Xenochrophis trianguligerus	+	+	+	+	+	+	-	LC
Striped Keelback Water Snake	Xenochrophis vittatus	-	-	+	+	-	+	-	LC
Pseudoxenodontidae (False Cobras)									
Blakeway's Blotch-necked Snake	Plagiopholis blakewayi	+	+	-	-	-	-	-	LC
Common Blotch-necked Snake	Plagiopholis nuchalis	+	+	-	-	-	-	-	LC
Baram False Cobra	Pseudoxenodon baramensis	-	-	-	-	E	-	-	LC
Javanese False Cobra	Pseudoxenodon inornatus	-	-	-	-	-	E	-	LC
Jacobson's False Cobra	Pseudoxenodon jacobsonii	-	-	-	E	-	-	-	DD
Large-eyed False Cobra	Pseudoxenodon macrops	+	+	+	-	-	-	-	LC
Elapidae (Cobras and Kraits, Coral and Sea Snakes)									
Himalayan Krait	Bungarus bungaroides	+	-	-	-	-	-	-	NE
Malayan Krait	Bungarus candidus	-	+	+	+	-	+	+	LC
Banded Krait	Bungarus fasciatus	+	+	+	+	+	+	-	NE
Red-headed Krait	Bungarus flaviceps	+	+	+	+	+	+	-	LC
Splendid Krait	Bungarus magnimaculatus	E	-	-	-	-	-	-	LC
Many-banded Krait	Bungarus multicinctus	+	-	-	-	-	-	-	LC
Black Krait	Bungarus niger	+	-	-	-	-	-	-	NE
Wanghaoting's Krait	Bungarus wanghaotingi	+	-	-	-	-	-	-	LC
Blue Coral Snake	Calliophis bivirgatus	+	+	+	+	+	+	-	LC
Spotted Coral Snake	Calliophis gracilis	-	+	+	+	-	-	-	DD
Malayan Striped Coral Snake	Calliophis intestinalis	-	+	+	+	+	+	-	LC
Speckled Coral Snake	Calliophis maculiceps	+	+	+	-	-	-	-	LC
Aagaard's Sea Snake	Hydrophis aagaardi	-	+	+	+	+	-	-	NE
Annandale's Sea Snake	Hydrophis annandalei	-	+	+	+	+	-	-	NE
Anomalous Sea Snake	Hydrophis anomalus	-	+	+	+	+	+	-	DD
Black-headed Sea Snake	Hydrophis atriceps	-	+	?	-	-	-	-	LC
Captain Belcher's Sea Snake	Hydrophis belcheri	-	+	-	-	+	+	-	DD
Two-wattled Sea Snake	Hydrophis bituberculatus	-	+	-	-	-	-	-	DD
Rajah Brook's Sea Snake	Hydrophis brookii	-	+	+	+	+	+	-	LC

Common English Name	Scientific Name	Myanmar	Thailand	Peninsular Malaysia	Sumatra	Borneo	Java	Bali	Global Status
Blue-grey Sea Snake	*Hydrophis caerulescens*	+	+	+	+	+	–	–	LC
Cantor's Sea Snake	*Hydrophis cantoris*	+	+	+	–	–	–	–	DD
Short Sea Snake	*Hydrophis curtus*	+	+	+	–	–	–	–	LC
Annulated Sea Snake	*Hydrophis cyanocinctus*	–	+	+	–	+	–	–	LC
Beaded Sea Snake	*Hydrophis eydouxii*	–	+	+	+	+	+	–	LC
Banded Sea Snake	*Hydrophis fasciatus*	+	+	+	–	–	+	–	LC
Narrow-headed Sea Snake	*Hydrophis gracilis*	+	+	+	+	+	–	–	LC
Plain Sea Snake	*Hydrophis inornatus*	–	+	+	–	–	+	–	DD
Saddle-backed Sea Snake	*Hydrophis jerdoni*	+	+	+	+	+	–	–	LC
Kloss's Sea Snake	*Hydrophis klossi*	–	+	+	+	+	–	–	DD
Lambert's Sea Snake	*Hydrophis lamberti*	–	+	+	–	–	–	–	LC
Persian Gulf Sea Snake	*Hydrophis lapemoides*	–	+	+	–	–	–	–	LC
Lesser Dusky Sea Snake	*Hydrophis melanosoma*	–	+	+	+	+	–	–	DD
Black-ringed Sea Snake	*Hydrophis nigrocinctus*	+	–	–	–	–	–	–	DD
Obscure-patterned Sea Snake	*Hydrophis obscurus*	+	–	–	–	–	–	–	LC
Ornate Sea Snake	*Hydrophis ornatus*	+	+	+	+	+	–	–	LC
Horned Sea Snake	*Hydrophis peronii*	–	+	+	–	–	–	–	LC
Yellow-bellied Sea Snake	*Hydrophis platura*	+	+	+	+	+	+	–	LC
Beaked Sea Snake	*Hydrophis schistosa*	+	+	+	+	+	–	–	LC
Sibau River Sea Snake	*Hydrophis sibauensis*	–	–	–	–	E	–	–	DD
Spiral Sea Snake	*Hydrophis spiralis*	+	+	+	+	+	–	–	LC
Stokes's Sea Snake	*Hydrophis stokesii*	–	+	+	+	+	–	–	LC
Narrow-necked Sea Snake	*Hydrophis stricticollis*	+	–	–	–	–	–	–	DD
Garlanded Sea Snake	*Hydrophis torquatus*	–	+	+	+	+	–	–	DD
Grey Sea Snake	*Hydrophis viperina*	+	+	+	–	+	–	–	LC
Yellow-lipped Sea Krait	*Laticauda colubrina*	+	+	+	+	+	+	+	LC
Large-scaled Sea Krait	*Laticauda laticaudata*	+	+	+	+	+	+	+	LC
Monocled Cobra	*Naja kaouthia*	+	+	+	–	–	–	–	LC
Mandalay Cobra	*Naja mandalayensis*	E	–	–	–	–	–	–	VU
Indo-Chinese Spitting Cobra	*Naja siamensis*	–	+	–	–	–	–	–	VU
Equatorial Spitting Cobra	*Naja sputatrix*	–	–	–	–	–	+	+	LC
Sumatran Cobra	*Naja sumatrana*	–	+	+	+	+	–	–	LC
King Cobra	*Ophiophagus hannah*	+	+	+	+	+	+	+	VU
MacClelland's Coral Snake	*Sinomicrurus macclellandi*	+	+	–	–	–	–	–	NE
Homalopsidae (Puff-faced Water Snakes)									
Keel-bellied Water Snake	*Bitia hydroides*	+	+	+	–	–	–	–	LC
Yellow-banded Mangrove Snake	*Cantoria violacea*	+	+	+	+	+	–	–	LC
Dog-faced Water Snake	*Cerberus rynchops*	+	+	–	–	–	–	–	LC

Common English Name	Scientific Name	Myanmar	Thailand	Peninsular Malaysia	Sumatra	Borneo	Java	Bali	Global Status
Schneider's Dog-faced Water Snake	Cerberus schneiderii	–	+	+	+	+	+	+	NE
Chan-ard's Water Snake	Enhydris chanardi	–	E	–	–	–	–	–	DD
Rainbow Water Snake	Enhydris enhydris	+	+	+	+	+	+	–	LC
Jagor's Water Snake	Enhydris jagorii	–	E	–	–	–	–	–	DD
Indo-Chinese Water Snake	Enhydris subtaeniata	–	+	–	–	–	–	–	LC
Tentacled Snake	Erpeton tentaculatus	–	+	–	–	–	–	–	LC
Siebold's Water Snake	Ferania sieboldii	?	–	–	–	–	–	–	LC
Crab-eating Mangrove Snake	Fordonia leucobalia	+	+	+	+	+	+	–	LC
Glossy Marsh Snake	Gerarda prevostiana	+	+	+	–	–	–	–	LC
Spotted Water Snake	Gyiophis maculosa	E	–	–	–	–	–	–	DD
Voris's Water Snake	Gyiophis vorisi	E	–	–	–	–	–	–	EN
Marquis Doria's Water Snake	Homalophis doriae	–	–	–	–	E	–	–	LC
Gyi's Water Snake	Homalophis gyii	–	–	–	–	E	–	–	DD
Puff-faced Water Snake	Homalopsis buccata	–	+	+	+	+	–	–	LC
Cox's Water Snake	Homalopsis mereljcoxi	–	E	–	–	–	–	–	NE
Black-bellied Water Snake	Homalopsis nigroventralis	–	E	–	–	–	–	–	NE
Martaban Water Snake	Homalopsis semizonata	E	–	–	–	–	–	–	NE
Grey Water Snake	Hypsiscopus plumbea	+	+	+	+	+	+	+	LC
Pahang	Kualatahan pahangensis	–	–	E	–	–	–	–	DD
Yellow-banded Water Snake	Miralia alternans	–	–	–	+	+	–	–	DD
Spotted Water Snake	Phytolopsis punctata	–	–	+	+	+	–	–	DD
Indian Water Snake	Raclitia indica	–	–	E	–	–	–	–	DD
Bocourt's Water Snake	Subsessor bocourti	–	+	+	–	–	–	–	LC
White-spotted Water Snake	Sumatranus albomaculata	–	–	–	E	–	–	–	DD
Pareidae (Slug-eating Snakes)									
Blunt-headed Slug Snake	Aplopeltura boa	+	+	+	+	+	+	–	LC
Bornean Slug Snake	Asthenodipsas borneensis	–	–	–	–	+	–	–	NE
Sabah Slug Snake	Asthenodipsas jamilinaisi	–	–	–	–	+	–	–	NE
Smooth Slug Snake	Asthenodipsas laevis	–	+	+	+	+	+	–	LC
Malayan Slug Snake	Asthenodipsas malaccanus	–	+	+	+	+	–	–	LC
Mountain Slug-eating Snake	Asthenodipsas vertebralis	–	–	E	–	–	–	–	LC
Keeled Slug-eating Snake	Pareas carinatus	+	+	+	+	+	+	+	LC
Twin Slug-eating Snake	Pareas geminatus	–	E	–	–	–	–	–	NE
Hampton's Slug-eating Snake	Pareas hamptoni	+	+	–	–	–	–	–	LC
White-spotted Slug-eating Snake	Pareas margaritophorus	+	+	+	–	–	–	–	LC
White-spotted Slug-eating Snake	Pareas monticola	+	–	–	–	–	–	–	NE
Barred Slug-eating Snake	Pareas nuchalis	–	–	–	–	E	–	–	LC
Vindum's Slug-eating Snake	Pareas vindumi	+	–	–	–	–	–	–	NE

Common English Name	Scientific Name	Myanmar	Thailand	Peninsular Malaysia	Sumatra	Borneo	Java	Bali	Global Status
Viperidae (Vipers and Pit Vipers)									
Fea's Viper	*Azemiops feae*	+	–	–	–	–	–	–	LC
Malayan Pit Viper	*Calloselasma rhodostoma*	–	+	+	–	–	+	–	LC
White-lipped Pit Viper	*Cryptelytrops albolabris*	+	+	–	–	+	–	–	LC
Cardamom Mountains Pit Viper	*Cryptelytrops cardomomensis*	–	+	–	–	–	–	–	LC
Spot-tailed Pit Viper	*Cryptelytrops erythrurus*	+	–	–	–	–	–	–	LC
Lesser Sunda White-lipped Pit Viper	*Cryptelytrops insularis*	–	–	–	–	–	–	+	LC
Kanburi Pit Viper	*Cryptelytrops kanburiensis*	–	E	–	–	–	–	–	EN
Large-eyed Pit Viper	*Cryptelytrops macrops*	–	+	–	–	–	–	–	LC
Mangrove Pit Viper	*Cryptelytrops purpureomaculatus*	+	+	+	–	–	–	–	LC
Beautiful Pit Viper	*Cryptelytrops venustus*	–	+	+	–	–	–	–	NE
Eastern Russell's Viper	*Daboia siamensis*	+	+	–	–	–	+	–	LC
Kinabalu Brown Pit Viper	*Garthius chaseni*	–	–	–	–	E	–	–	LC
Malayan Brown Pit Viper	*Ovophis convictus*	–	+	+	+	–	–	–	LC
Blotched Pit Viper	*Ovophis monticola*	+	–	–	–	–	–	–	LC
Gunalen's Pit Viper	*Parias gunaleni*	–	–	–	+	–	–	–	NE
Hagen's Green Pit Viper	*Parias hageni*	–	+	+	+	–	–	–	LC
Kinabalu Green Pit Viper	*Parias malcolmi*	–	–	–	–	E	–	–	NT
Sumatran Pit Viper	*Parias sumatranus*	–	+	+	+	+	–	–	LC
Sumatran Green Pit Viper	*Popeia barati*	–	–	–	+	–	–	–	LC
Pulau Tioman Pit Viper	*Popeia buniana*	–	–	+	–	–	–	–	EN
Thai Peninsula Pit Viper	*Popeia fucata*	+	+	+	–	–	–	–	LC
Cameron Highlands Pit Viper	*Popeia nebularis*	–	–	E	–	–	–	–	VU
Phuket Pit Viper	*Popeia phuketensis*	–	E	–	–	–	–	–	NE
Pope's Pit Viper	*Popeia popeiorum*	+	+	–	–	–	–	–	LC
Sabah Green Pit Viper	*Popeia sabahi*	–	–	–	–	E	–	–	LC
Lake Toba Pit Viper	*Popeia toba*	–	–	–	E	–	–	–	DD
Jerdon's Pit Viper	*Protobothrops jerdonii*	+	–	–	–	–	–	–	LC
Kaulback's Pit Viper	*Protobothrops kaulbacki*	+	–	–	–	–	–	–	LC
Omkoi Lance-headed Pit Viper	*Protobothrops kelomohy*	–	+	–	–	–	–	–	NE
Brown-spotted Pit Viper	*Protobothrops mucrosquamatus*	+	–	–	–	–	–	–	LC
Sumatran Palm Pit Viper	*Trimeresurus andalasensis*	–	–	–	E	–	–	–	LC
Bornean Palm Pit Viper	*Trimeresurus borneensis*	–	–	–	–	E	–	–	LC
Brongersma's Palm Pit Viper	*Trimeresurus brongersmai*	–	–	–	E	–	–	–	NT
Javanese Palm Pit Viper	*Trimeresurus puniceus*	–	–	–	+	–	+	–	LC
Wirot's Palm Pit Viper	*Trimeresurus wiroti*	–	+	+	–	–	–	–	LC

Common English Name	Scientific Name	Myanmar	Thailand	Peninsular Malaysia	Sumatra	Borneo	Java	Bali	Global Status
Bornean Keeled Green Pit Viper	*Tropidolaemus subannulatus*	–	–	–	–	+	–	–	LC
Wagler's Keeled Green Pit Viper	*Tropidolaemus wagleri*	–	+	+	+	–	–	–	LC
Gumprecht's Pit Viper	*Viridovipera gumprechti*	+	+	–	–	–	–	–	LC
Medo Pit Viper	*Viridovipera medoensis*	+	–	–	–	–	–	–	DD
Stejneger's Pit Viper	*Viridovipera stejnegeri*	+	–	–	–	–	–	–	LC
Vogel's Pit Viper	*Viridovipera vogeli*	–	+	–	–	–	–	–	LC
Xenodermatidae (Strange-skinned Snakes)									
Laos Forest Snake	*Parafimbrios lao*	–	+	–	–	–	–	–	NE
Stoliczka's Stream Snake	*Stoliczkia borneensis*	–	–	–	–	E	–	–	LC
Rough-backed Litter Snake	*Xenodermus javanicus*	+	+	+	+	+	+	–	LC
Xenophidiidae (Spine-jawed Snakes)									
Bornean Spiny-jawed Snake	*Xenophidion acanthognathus*	–	–	–	–	E	–	–	DD
Schäfer's Spiny-jawed Snake	*Xenophidion schaeferi*	–	–	E	–	–	–	–	DD
Typhlopidae (Blind Snakes)									
Burmese Blind Snake	*Argyrophis bothriorhynchus*	+	–	?	–	–	–	–	DD
Diard's Blind Snake	*Argyrophis diardii*	+	+	–	–	–	–	–	LC
Pitted Snout Blind Snake	*Argyrophis hypsobothrius*	–	–	–	E	–	–	–	DD
Klemmer's Blind Snake	*Argyrophis klemmeri*	–	E	–	–	–	–	–	DD
Müller's Blind Snake	*Argyrophis muelleri*	+	+	+	+	+	–	–	LC
Oates's Blind Snake	*Argyrophis oatesii*	+	–	–	–	–	–	–	DD
Roxane's Blind Snake	*Argyrophis roxaneae*	–	E	–	–	–	–	–	DD
Thai Blind Snake	*Argyrophis siamensis*	–	+	–	–	–	–	–	DD
Trang Blind Snake	*Argyrophis trangensis*	–	E	–	–	–	–	–	DD
White-headed Blind Snake	*Indotyphlops albiceps*	+	+	+	–	–	–	–	LC
Brahminy Blind Snake	*Indotyphlops braminus*	+	+	+	+	+	+	+	NE
Jerdon's Blind Snake	*Indotyphlops jerdoni*	+	–	–	–	–	–	–	NE
Ozaki's Blind Snake	*Indotyphlops mollyozakiae*	–	E	–	–	–	–	–	DD
Large Blind Snake	*Indotyphlops porrectus*	+	–	–	–	–	–	–	NE
Koekkoek's Blind Snake	*Malayotyphlops koekkoeki*	–	–	–	–	E	–	–	DD
Lined Blind Snake	*Ramphotyphlops lineatus*	–	+	+	+	+	+	–	LC
Lorenz's Blind Snake	*Ramphotyphlops lorenzi*	–	–	–	–	E	–	–	DD
Lorenz's Blind Snake	*Ramphotyphlops olivaceus*	–	–	–	–	+	–	–	NE
Gerrhopylidae (Indo-Malayan Blind Snakes)									
Black Blind Snake	*Gerrhopilus ater*	–	–	–	–	+	–	+	NE
Javanese Blind Snake	*Gerrhopilus bisubocularis*	–	–	–	–	–	E	–	DD
Flower's Blind Snake	*Gerrhopilus floweri*	–	E	–	–	–	–	–	DD

Auliya, M. (2006). *Taxonomy, Life History and Conservation of Giant Reptiles in West Kalimantan (Indonesian Borneo)*. Natur und Tier Verlag GmbH.

Chan-ard, T., Grossmann, W., Gumprecht, A. and Schulz, K.-D. (1999). *Amphibians and Reptiles of Peninsular Malaysia and Thailand: An Illustrated Checklist/Amphibien und Reptilien der Halbinsel Malaysia und Thailands: Eine illustrierte Checkliste*. Bushmaster Publications.

Cox, M.J., Van Dijk, P.P., Nabhitabhata, J. and Thirakhupt, K. (1998). *A Photographic Guide to Snakes and Other Reptiles of Peninsular Malaysia, Singapore and Thailand*. New Holland Publishers (UK) Ltd.

Das, I. (2006). *A Photographic Guide to the Snakes and Other Reptiles of Borneo*. New Holland Publishers (UK) Ltd.

Das, I. (2010). *A Field Guide to the Reptiles of South-east Asia*. New Holland Publishers (UK) Ltd.

David, P. and Vogel, G. (1996). *The Snakes of Sumatra. An Annotated Checklist and Key with Natural History Notes*. Edition Chimaira.

de Lang, R. and Vogel, G. (2005). *The Snakes of Sulawesi. A Field Guide to the Land Snakes of Sulawesi with Identification Keys*. Edition Chimaira.

de Rooij, N. (1917). *The Reptiles of the Indo-Australian Archipelago. II. Ophidia*. E.J. Brill.

Grismer, L.L. (2011). *Field Guide to the Amphibians and Reptiles of the Seribuat Archipelago (Peninsular Malaysia)*. Edition Chimaira.

Gumprecht, A., Tillack, F., Orlov, N.L., Captain, A. and Ryabov, S. (2004). *Asian Pitvipers*. GeitjeBooks.

Jaafar, I., Sah, S.A.M., Ahmad, N., Chan, K.O. and Akil, M.A.M.M. (2008). *The Common Amphibians and Reptiles of Penang Island*. State Forestry Department of Penang.

Kurniati, H. (2003). *Amphibians and Reptiles of Gunung Halimun National Park, West Java, Indonesia (Frogs, Lizards and Snakes). An Illustrated Guide Book*. Research Center for Biology – LIPI.

Leviton, A. E., Wogan, G., Koo, M., Zug, G. R. and Vindum, J. V. (2003). *The dangerously venomous snakes of Myanmar*. Illustrated checklist with keys. Proceedings of the California Academy of Sciences 54(24):407–462.

Lim, K.K.P. and Lim, F.L.K. (2002). *A Guide to the Amphibians and Reptiles of Singapore*. Revised edn. Singapore Science Centre.

McKay, L. (2006). *A Field Guide to the Reptiles and Amphibians of Bali*. Krieger Publishing.

Malkmus, R., Manthey, U., Vogel, G., Hoffmann, P. and Kosuch, J. (2002). *Amphibians and Reptiles of Mount Kinabalu (North Borneo)*. Koeltz Scientific Books.

Manthey, U. and Grossmann, W. (1997). *Amphibien und Reptilien Südostasiens*. Natur und Tier.

O'Shea, M. (2007). *Boas and Pythons of the World*. New Holland Publishers (UK) Ltd.

Schulz, K.-D. (1996). *A Monograph of the Colubrid Snakes of the Genus* Elaphe *Fitzinger*. Koeltz Scientific Books.

Smith, M.A. (1943). *The Fauna of British India, Ceylon and Burma, Including the Whole of the Indo-Chinese Region. Vol. III*. Serpentes. Taylor & Francis.

Somaweera, R. (2017). *A Naturalist's Guide to the Reptiles and Amphibians of Bali*. John Beaufoy Publishing.

Stuebing, R.B. and Inger, R.F. (1999). *A Field Guide to the Snakes of Borneo*. Natural History Publications (Borneo) Sdn Bhd.

Tweedie, M.W.F. (1983). *The Snakes of Malaya*. 3rd edn. Singapore National Printers.

Vogel, G. (2006). *Venomous Snakes of Asia/Giftschlangen Asiens*. Terralog 15. Edition Chimaira.

World Health Organization. 2016. *WHO Guidelines for the Production, Control and Regulation of Snake Antivenom Immunoglobulins*. WHO Press, Geneva.

I thank John Beaufoy for the invitation to prepare this book, and I am also grateful to Ken Scriven for the initial idea and for encouragement. I remain indebted to the Thai consultant of this work, the late Jonathan Murray for his support, including organising photos of additional species and for comments on format and contents, and to the book's copy editor, Susi Bailey, and commissioning editor, Rosemary Wilkinson.

Manuscript preparation was supported by Universiti Malaysia Sarawak, and I thank our director, Andrew Alek Tuen, and other colleagues and students, for their support. For generous sharing of data, including papers and unpublished information, I remain grateful to the following colleagues: Kraig Adler, Ishan Agarwal, Natalia Ananjeva, Umilaela Arifin, E. Nicholas Arnold, Kurt Auffenberg, the late Walter Auffenberg, Mark Auliya, Christopher Austin, Raul Bain, David Barker, Aaron M. Bauer, David Bickford, Vladimir Bobrov, Wolfgang Böhme, Rafe Brown, John Cadle, Ashok Captain, Datuk Chan Chew Lun, Lawan Chanhome, Joseph Charles, Michael Cotta, Merrel Jack Cox, Jennifer Daltry, the late Ilya Darevsky, Abhijit Das, Patrick David, Stuart Davies, Geoff Davison, Maximilian Dehling, Arvin C. Diesmos, Cheong-Hoong Diong, Julian Dring, David S. Edwards, Kelvin Egay, Linda Ford, Maren Gaulke, Frank Glaw, David Gower, Ulmar Grafe, Allen Greer, Jesse Grismer, Lee Grismer, Wolfgang Grossmann, Andreas Gumprecht, Rainer Günther, Alexander Haas, Jakob Hallermann, Amir Hamidy, James Hanken, Harold Heatwole, Kelvin Kueh Boon Hee, Stefan Hertwig, Ronald Heyer, Tsutomu Hikida, Marinus Hoogmoed, Jaafar Ibrahim, Ivan Ineich, Robert F. Inger, Djoko Iskandar, Jiang Jian-Ping, Ulrich Joger, David Jones, Vladimir Kharin, André Koch, Gunther Köhler, Maklarin Lakim, Charles Leh, Tzi Ming Leong, Alan Leviton, Harvey Lillywhite, Lim Boo Liat, Lim Chan Koon, Kelvin K.P. Lim, Aaron Lobo, Colin McCarthy, Jimmy McGuire, Steven Mahony, Anita Malhotra, Edmond Malnate, Ulrich Manthey, Joseph Martinez, the late Sherman Minton, Viral Mistry, G. Mumpuni, John Murphy, Jonathan Murray, the late Jarujin Nabhitabhata, Peter K.L. Ng, Truong Nguyen, Samhan Nyawa, Ong Jia Jet, Chan Kin Onn, Nikolai Orlov, Mark O'Shea, Hidetoshi Ota, Olivier S.G. Pauwels, Pui Yong Min, Ding Qi Rao, Arne Rasmussen, the late Jens Rasmussen, José Rosado, Gerold Schipper, Klaus-Dieter Schulz, Saibal Sengupta, Chris Shepard, Rick Shine, Irwan Sidik, the late Joseph Slowinski, Hobart Smith, Robert Stuebing, Bryan Stuart, Jeet Sukumaran, Mona Octavia Sulai, Montri Sumontha, Tan Heok Hui, Alexander Teynié, Kumthorn Thirakhupt, Frank Tillack, Oswald Braken Tissen, Michihisa Toriba, Ngo Van Tri, Nguyen Truong, Andrew Alek Tuen, the late Garth Underwood, Peter Paul van Dijk, Johan van Rooijen, Miguel Vences, Jens Vindum, Gernot Vogel, Harold Voris, Van Wallach, David Warrell, Romulus Whitaker, Antony Whitten, Guin Wogan, Perry Wood, Wolfgang Wüster, Norsham Yaakob, Paul Yambun, Yong Hoi Sen, Timothy Youmans, Er-Mi Zhao, Thomas Ziegler, Nikolay Zinenko and George Zug. For any I may have inadvertently left out, I offer my apologies.

I am grateful to a number of permitting bodies, including the Sarawak Forest Department, the Sarawak Forestry Corporation, Sabah Parks, PERHILITAN, Brunei Museums Department and the Brunei Forestry Department for research permits over the years. Fieldwork was supported by grants from Universiti Malaysia Sarawak (120(98)(9), 192(99)(46), 1/26/303/2002(40), 01/59/376/2003113); Intensification of Research in Priority Areas Grant, Malaysia (08-02-09-10007-EA0001); Fundamental Research Grant, Ministry of Higher Education, Malaysia (FRG/06(10)667/2007(32); Fundamental Research Grant, Ministry of Higher Education, Malaysia (FRGS/07(04)787/2010(68); and Shell Chair (SRC/05/2010[01]), administered by the Institute of Biodiversity and Environmental Conservation, Universiti Malaysia Sarawak. Additional support was provided by the United Nations Development Programme/Global Environmental Facility (UNDP/GEF Project MAL/99/G31); DANCED/Support for Wildlife Management Plan Implementation Project; Ramsar Center Japan; Forestry Department, Ministry of Industry and Primary Resources, Brunei Darussalam; Volkswagen-Stiftung (1/79 405); the Mohamed bin Zayed Species Conservation Fund (L18403 101 00mBZ); Conservation International; and the IUCN Amphibian Specialist Group. Conservation International supported my attendance at the IUCN Red List meeting in Beijing in 2011.

The majority of images used in this book are mine, taken during fieldwork and, more rarely, at captive

▪ Acknowledgements ▪

facilities. The earlier (pre-2005) images were taken using slide transparency film with Nikon equipment (initially, a F801s camera, and subsequently, an F5 camera and Micro-Nikkor 105mm lens, illuminated with one or two Speedlight flash systems), and scanned using a CanoScan slide scanner. Images from around 2005 were taken mostly in a digital format (Nikon D70 camera and 105mm Micro-Nikkor lens, and illuminated with a SB800 Speedlight flash). Additional images were generously provided by Mark Auliya, Ashok Captain, Susan Clark, Michael Cota, Abhijit Das, Maximilian Dehling, Lee Grismer, H.T. Lalremsanga, Hla Tun, Kirati Kunya, Ron Lilley, Kelvin K.P. Lim, Norman Lim, Dong Lin, Aaron Lobo, Ulrich Manthey, Viral Mistry, Jonathan Murray, Manoj Nair, Chan Kin Onn, Nikolai Orlov, Pui Yong Min, Klaus-Dieter Schulz, Jeet Sukumaran, Montri Sumontha, Tan Heok Hui, Alexander Teynié, Frank Tillack, Nguyen Truong, Gernot Vogel, Guin Wogan, Wolfgang Wüster, Thomas Ziegler and George Zug.

Images of additional species in the second edition were supplied by Kurt Orion, Khaldun Ismail, Vincent Eng Wah, Xavier Fenoy, Björn Lardner, Rudi Rahadian and Neil Rowntree. I thank Vincent Teo Eng Wah for comments.

For the third edition, images of additional species were provided by Taksa Vasaruchapong, Shavez Cheema, Peter Giessler, Tim Hodges, Mistar Kamsi, Nikolay Poyarkov, Evan Quah, Montri Sumontha, Vincent Eng Wah Teo, Vitaly Trounov and Jens Vindum.

Photo Credits
Main descriptions: photos are denoted by a page number followed by t (top), m (middle), b (bottom), l (left) or r (right), if relevant.

Mark Auliya: 17(b), 79(t), 92(t), 113(t), 125(b). Ashok Captain/The Lisus: 23(t), 50(b), 61(b), 75(b), 84(b), 94(b), 149(b). Chan Kin Onn: 127(b). Shavez Cheema: 34(t). Sue Clark: 49(t). Michael Cota: 54(t), 70(t), 138(b). Abhijit Das: 24(b), 27(t), 28(t), 28(b), 38(t), 39(b), 42(t), 43(b), 46(b), 55(b), 58(t), 65(t), 74(b), 75(t), 76(t), 81, 85(t), 85(b), 87(t), 87(b), 88(t), 90(b), 92(b), 94(t), 95(b), 98(b), 101(b), 115(b), 123(t), 136(t), 144(b), 144(t). Indraneil Das: 5, 6(t), 6(b), 10, 11, 12(t), 12(b), 14(t), 14(b), 15, 16(t), 16(b) 17(t), 18, 20(t), 21, 22(t), 24(t), 24(b), 25(t), 25(b), 26(b), 29(t), 30(t), 30(b), 32(t), 33(t), 33(b), 34(b), 35(t), 35(b), 36(t), 37, 38(t), 40(t), 41(t), 41(b), 44(b), 45, 46(t), 47(b), 48(t), 49(b), 51(b-l), 51(b-r), 52(t), 52(b), 54(b), 63(b), 65(b), 67(t), 68(t), 68(b), 69(b), 71(t), 71(b), 72(t), 72(b), 73(t), 73(b), 74(t), 76(b), 78(t), 79(b), 83(b), 84(t), 86(b), 88(b), 89(t), 90(t), 91(t), 91(b), 93(t), 95(t), 97(b), 98(t), 99, 103(b), 105(b), 106(t), 106(b), 107(b), 108(t), 111(t), 112(b), 115(t), 117(t), 117(b), 118(t), 121(t), 121(b), 122(b), 123(t), 124(t), 124(b), 125(t), 127(t), 129(t), 129(b), 130(t), 131(b), 133(t), 133(b), 135(t), 135(b), 137(b), 138(t), 139(t), 140(t), 140(b), 141(t), 141(b), 142(b), 143(t), 144(b), 146(b), 148(t), 148(b), 149(t), 150(b), 151(t), 151(b), 153, 154(t), 155(t), 155(b). Maximilian Dehling: 29(b), 62(t), 136(b). Dong Lin/California Academy of Sciences: 70(b), 120. Xavier Fenoy: 100(t), 100(b). Peter Giessler: 32(b), 154(b). Lee Grismer: 31(b), 43(t), 66(t), 107(t), 131(t). Hla Tun/California Academy of Sciences: 64(b), 96(b). Tim Hodges: 31(t), 78(b). Mistar Kamsi: 128(b). Kirati Kunya: 89(b). H.T. Lalremsanga: 27(b), 51(t), 77. Björn Lardner: 156. Ron Lilley: 63(t), 103(t). Kelvin Kok Peng Lim: 20(b), 53(t), 58(b), 86(t). Aaron Lobo: 108(t), 108(b), 109(t), 110(t), 112(t), 113(b), 114. Ulrich Manthey: 42(b). Viral Mistry: 102. Jonathan Murray: 104(b). Pui Yong Min: 130(b), 146(m). Manoj Nair: 105(t). Nikolai Orlov: 116(t), 118(t), 119. Nikolay Poyarkov: 44(t), 126(t-l), 126(t-r). Evan Quah: 38(b), 59(t), 60(t). Rudi Rahadian: 101(t). Phil Round: 83(t). Neil Rowntree: 152. Klaus-Dieter Schulz: 22(b), 26(t), 36(b), 47(t), 60(b), 134. Jeet Sukumaran: 104(t). Montri Sumontha: 56(b), 66(b), 82(t), 110(b), 122(t), 143(b). Tan Heok Hui: 111(b). Alexandre Teynié: 55(b), 56(t). Vincent Eng Wah Teo: 59(b). Frank Tillack: 57(b), 139(b), 142(t), 147(b). Vitaly Trounov: 128(t). Nguyen Quang Truong: 82(b), 96(b). Taksa Vasaruchapong: 53(b), 57(t), 145(b). Jens Vindum: 13, 126(b). Gernot Vogel: 19(t), 19(b), 40(b), 48(b), 61(t), 64(t), 67(b), 93(b), 97(t), 132(b), 137(b), 146(t), 150(t). Wolfgang Wüster: 50(t), 63(b), 80, 116(b), 147(t). Thomas Ziegler: 132(t). George R. Zug: 69(t).